Reading Ability

Reading Ability

Charles A. Perfetti

University of Pittsburgh

New York • Oxford University Press
1985

Oxford University Press, Walton Street, Oxford OX2 6DP

London   New York   Toronto
Delhi   Bombay   Calcutta   Madras   Karachi
Kuala Lumpur   Singapore   Hong Kong   Tokyo
Nairobi   Dar es Salaam   Cape Town
Melbourne   Auckland

and associated companies in
Beirut   Berlin   Ibadan   Mexico City   Nicosia

Library of Congress Cataloging in Publication Data
Perfetti, Charles A.
Reading ability.
Includes index.
1. Reading.   I. Title.
LB1050.P385   1984   428.4   84-25413
ISBN 0-19-503501-1

Printing (last digit): 9 8 7 6 5 4 3 2 1

Printed in the United States of America

To my mother, Oma Hager Perfetti,
and to the memory of my father, Otto Perfetti;
to Lisa and Michael.

Preface

Acquiring competence in reading proves to be an easy matter for some children, whereas for others it is a problem. Similarly, adults vary in their ability to read with comprehension. This book is concerned with the following general question: How can we explain such differences in reading ability?

There are many scientific and practical issues arising from this question. On the practical side, we want to know how beginning reading can be taught to reduce the number of failures. And we want to know how to improve the reading skills of children and adults who have problems beyond beginning reading. On the scientific side, we want to know about how reading works, how it depends on a script system, how words are identified, how sentences and texts are comprehended, how reading is related to spoken language processes, what causes dyslexia, and so forth. All of these and many other issues are addressed in this book, the guiding principle being that the practical questions have to be linked with the scientific ones.

The approach I take to these questions reflects years of research. It is the approach of someone who is a cognitive psychologist, with a special focus on psycholinguistics. This book, in a sense, represents a psycholinguistic approach to the study of reading ability. However, I am reluctant to give it this label because "psycholinguistic" has long been applied to an approach to reading (Goodman, 1967) that differs in important ways from my own. Nevertheless, my approach does emphasize the language dependency of reading. It acknowledges the importance of visual processes in reading and it examines the ways in which reading and spoken language are different. In the final analysis, it is my conclusion that the language-related aspects of reading seem to be most important in explaining differences in reading ability—differences that emerge in the acquisition of reading skills, in comprehension, and in dyslexia.

   A few general features of the book are worth noting. The main practical point to be stressed is that while it provides a comprehensive discussion of reading ability, it does present certain theoretical arguments. This
is reflected in the second section, which gives an account of verbal efficiency theory as a framework for understanding reading ability. The theory itself is explained in the sixth chapter, with the fifth and seventh
chapters providing some of the evidence that supports it. The exposition
is purposely uncluttered in the hope that the reportage and application
of the data to the issues will be clear to students and other readers who
are not researchers. Thus, although much of the research on individual
differences in reading ability would further support the theoretical arguments presented here, or refine them in a particular direction, these
chapters draw mainly on my own research to illustrate the kinds of evidence that bear on the theoretical framework. I focused on my work both
because it is internally consistent in its sampling methods and its definitions of reading ability and because the different approaches to ability
groupings make comparisons with other research difficult. (Of course, I
also claim an author's privilege to report on his or her own research.)
   This book, then, is not a survey of theories and research in reading
ability. It does, however, treat the broad range of issues in reading and
reading acquisition and thus could be a suitable textbook for a course in
the psychology of reading. The theoretical claims are embedded in a very
broad discussion of the processes of reading, mainly in the first section.
The second chapter, on lexical access, and the third, on comprehension,
reflect, within the cognitive framework set out in the first chapter, the
basic issues in these two major components of reading. The fourth chapter discusses the issues of speech processes in reading and presents some
theoretical suggestions as to how such processes can be understood.
   The book is also comprehensive in the scope of reading ability that is
covered. The eighth chapter discusses the development of reading ability
through adulthood and suggests some common processing factors that are
involved throughout the age range. The ninth chapter discusses dyslexia,
suggesting some continuity between ordinary low levels of reading ability and at least most dyslexias. The final two chapters deal with beginning reading and reading instruction. The tenth chapter provides a bal-
anced treatment of beginning reading and the eleventh suggests how some
topics discussed in earlier chapters can be applied to issues of reading
instructio. , both beginning reading and remedial instruction.
   Thus, the book attempts to provide a more complete account of verbal
efficiency theory than has been available in previous versions (e.g., Perfetti & Lesgold, 1977). In so doing, it both refers to and reports data, both
published and unpublished. But it does so in the context of a comprehensive range of reading ability issues, from instruction to dyslexia, from

beginning reading to adult reading, and from word recognition to comprehension. It is my hope that if this book makes a modest contribution to the growing theoretical literature on reading, it also serves to introduce anyone with an interest in reading to the cognitive analysis of reading skill. The theoretical arguments and the supporting research will be of interest to reading researchers and students. Moreover, I particularly hope that readers who have practical concerns, especially those who are involved with how to teach reading and with remediating reading disabilities, will find something of value in the following pages.

Pittsburgh   C.A.P.
July 1984

Acknowledgments

This book has been made possible thanks to the help of many people over a long period. It is a pleasure to be able to acknowledge those individuals whose contribution directly and indirectly made a difference in terms of the shape of the book's ideas and the realization of the book itself.

First I wish to acknowledge the intellectual and scholarly help of colleagues and students. I have been fortunate to be able to carry out research in an environment that fosters not merely collaboration but genuine encouragement and criticism within a community of shared interests. For this I am especially grateful to Robert Glaser and Lauren Resnick, the codirectors of the Learning Research and Development Center (LRDC) at the University of Pittsburgh. They are not only partly responsible for this nurturant environment but have been both supportive and constructive in challenging some of my ideas about reading.

A group of researchers at LRDC has had a significant influence on my work. Alan Lesgold provided much insight in our earlier collaborations on reading, especially in the areas of theory, metaphor, and irreverent cognizing. Whether he played Laurel to my Hardy, or vice versa, remains a matter for future scholarship to decide. In either case, his contributions to the ideas in this book have been enduring. I am likewise grateful to Isabel Beck, James Voss, Lauren Resnick, Steve Roth, Margaret Mc-Keown, and other members of the reading research group at LRDC. Isabel Beck's firsthand knowledge of how reading works in schools has had an influence on my attempts to apply verbal efficiency theory to instruction. Both her own research and our collaborative ventures have contributed much to some of the conclusions in this book. Jim Voss's important research on the role of knowledge in comprehensions has likewise had a significant influence on how I have dealt with reading comprehension.

The contributions of both Voss and Beck, as well as those of Lesgold and Resnick, are indeed quite obvious in many places. So, too, is the work of Richard Omanson, undertaken during his postdoctoral stay in Pittsburgh. Equally important has been the intellectual stimulation provided by these and other colleagues, too numerous to mention, whom I have had the good luck to work with over the years.

Among those who have made this work possible are graduate students, past and present, who have collaborated with me on reading research and have significantly helped to shape it. Sylvia Beverly, Leslie Leahy, Kathy Hammond, Deborah McCutchen, Steve Roth, Susan Goldman, Thomas Hogaboam, Mary Riley, and Natalie Berger carried out research that provides the empirical base for some of the conclusions that emerge in the book. Each of them has contributed at least one and, in some cases, several significant pieces of research on reading. Mary Beth Curtis, both as a student and after, carried out research that influenced my understanding of reading. I am grateful to Deborah McCutchen not only for research collaboration but for many helpful discussions about reading and language.

One constant and indispensable contributor to this research has been Laura Bell. She has assisted in most of the research referred to in this book. Her contributions have been invaluable. I am indebted to her for everything from experimental work of all kinds to a careful reading of the manuscript itself.

I am also grateful to the many school personnel who have kindly facilitated our research efforts over the years. These especially include the teachers, students, and principals of St. Joseph's (Mt. Oliver), St. James' (West End), and the South Side Catholic Schools (St. Adalbert's, St. Mathew's, and, formerly, St. Josephat's and St. Peter's), all in Pittsburgh.

Work on the book has benefited from many individual researchers in the field. I especially thank Keith Rayner for a thoughtful review of the entire manuscript and Richard Olson for some illuminating discussions concerning reading ability and for trying out an unpublished version of the manuscript in his class.

I began the writing of this book during the fall of 1982, when I was a visiting fellow at the Max-Planck-Institut für Psycholinguistik in Nijmegen. In fact, I completed first drafts of eight chapters during my stay in Nijmegen, and it is unlikely that I could have made such headway without the opportunity extended me by the institute. For this I am grateful to Willem Levelt and Wolfgang Klein, codirectors of the Max-Planck-Institut at that time. I am also grateful to Leo Noordman and to Witske Vonk for acting as hosts for my visit and for helpful discussions about reading comprehension. I also had stimulating discussions with many people both at the institute and at Nijmegen University, including Tom

Hudson, Annette deGroot, G. B. Flores d'Arcais, G. Seegers, J. Feenstra, Ann Cutler, Graham Richardson, Lorraine Tyler, Maya Hickmann, Annette Friederici—and especially Ulrich Glowalla, whose visit to the Max-Planck-Institut happily coincided with my own. Many of these discussions were helpful in the writing of the draft chapters. So, too, were discussions with Kees van den Bos in Groningen, H. Bouma and D. Bouwhuis in Eindhoven, and Heinz Mandl and his research colleagues in Tübingen during this same period.

I am also grateful to the National Institute of Education, which has supported most of my research on reading through its support of LRDC.

Finally, I want to acknowledge the important help in manuscript preparation that I received from several people. Shelley Reinhardt of Oxford University Press was a very helpful editor, especially at critical times in the life of the manuscript. Patricia Kelley created a number of appealing illustrations, and Connie Fabrizio helped produce a reference list. I especially want to acknowledge the fine manuscript preparation work of Jay Ritter and Sharon Vais. Jay transformed scrawls into first-draft text and Sharon did the same for later drafts. I am grateful to both.

Contents

PART ONE

Basic Reading Processes

CHAPTER ONE

Reading Ability: Cognitive and Psycholinguistic Perspectives

Reading is both simple and complex. It is, at the same time, both cognitively trivial and so difficult that failure at learning to read is common. Edmund Burke Huey (1908), who published the first comprehensive account of reading, could perceptively observe that a complete theory of reading would involve a description of "very many of the most intricate workings of the human mind" (p. 6). On the other hand, countless parents with no expertise in reading can boast of their preschool children's prodigious accomplishments in literacy. For example, Roger Schank (1982) begins his book on what artificial intelligence might tell us about reading by observing that he taught his daughter to read when she was four years old. Of course, reading achievements at an even younger age are possible. Just ask a parent.

However, not all children are so lucky. Many children do so poorly at reading that they are given a special category, "developmental dyslexia." Moreover, quite aside from specialized reading disabilities, there is continuing widespread concern about the ordinary failures of reading experienced by countless numbers of children who never earn the "dyslexic" label.

The issues of reading ability are not limited to children, however. Among adults, there are countless semiliterates. Moreover, even among college students, all of whom are literate, there is a wide range of talent. Some can read a textbook at 400 words per minute with good comprehension. Others plod through the same book at 100 words per minute and with no better comprehension than the faster students.

Thus, the range of reading talent is very broad and has several components. Huey was clearly correct about the "intricate workings of the human mind" being reflected in reading. It will not be possible to understand variations in reading talent without taking note of some of these intricacies. In short, an understanding of reading ability requires an understanding of reading. This means that we must know something about the components of reading.

COGNITIVE COMPONENTS OF READING

When a sentence such as (1) below is read, some of the components of reading are seen:

(1) *Marvin asked his brother to lend him five dollars. He refused.*

The first obvious component is that the words be "recognized." That is, *Marvin, asked, his*, etc., must be associated with familiar concepts represented in the reader's memory. These representations include names of persons (Marvin), general concepts *(ask, brother, lend, dollar)*, and referring devices, or anaphora *(he, his)*. In each case, they also include information about how the word is pronounced. Since these recognition processes involve access to mental representations of words, this general component of reading is referred to as *lexical access*.

The second general component is comprehension, which is not a single component at all, but a set of interrelated processes. In fact, it is reasonable to think of lexical access as one of these processes. However, the more salient comprehension components are those by which the reader builds a representation of text meaning. With a good representation of the text, the reader can answer such simple questions as: *(a)* What did Marvin do? *(b)* Whom did Marvin ask to lend him money? *(c)* How much money did Marvin get from his brother? Quite a few questions can be used to tap the representation of text meaning. Some, such as *(a)* and *(b)* tap a representation of text information explicitly in the text. Others, such as *(c)*, tap a representation that includes inferred information. Building the text representation, even for such a short text is as intricate in reality as it is simple in appearance. Think, for example, of the linguistic and practical knowledge that allows assignment of reference for *he* in the sentence *he refused.*

These intricate comprehension processes include *propositional encoding* and *propositional integration*, which are the relatively local text processes, and *text modeling*, which includes the processes of constructing the overall meaning of a text. The local comprehension processes are those that allow comprehension of even isolated sentences and sentence pairs. The text-modeling processes use the outcome of these local processes to build a larger representation of extended text.

Thus, the general account of reading processes includes components of lexical access and comprehension. The reader identifies words, taking account of their meaning and pronunciation as context and task demands require. Complex comprehension processes build on these recognition processes and perhaps even influence them in a limited way. These are the reading processes that are to be understood if we are to understand reading ability.

LANGUAGE AND READING

The central processes of reading are essentially mental operations on linguistic structures that begin with visual input. Although visual processes and linguistic processes have to be taken into account, the linguistic processes may turn out to be especially important for reading ability.

This is not to say that the visual processing of letter and word forms, or Chinese characters for that matter, is not important. Indeed, a significant part of becoming skilled in reading is the increase, with experience, of recognition for print patterns. Gibson and Levin (1975) emphasized the role of this kind of perceptual learning. The reader learns to discriminate among letter forms, then letter patterns. We can think of the reader with years of experience as having countless useful patterns represented in memory. These patterns are abstract—they account for TH and th irrespective of the exact shape of the characters—and they potentially include a wide range of units—single letters, letter patterns (th, ing), and whole words (the). It is conceivable that even longer words, provided they have been encountered often enough, can be represented as patterns in memory. Thus, normal reading includes an obvious visual component and one to which learning can apply.

The linguistic part of reading is emphasized for two reasons. The main reason is that reading includes both recognition of words and comprehension. Linguistic processes are heavily involved in both of these. Comprehension processes are essentially manipulations of linguistic objects rather than visual ones. Word recognition processes are essentially translations of visual objects into linguistic symbols used in these manipulations. A secondary reason for the linguistic emphasis follows from the first: A theory of reading ability must take account of the linguistic aspects of reading because differences in reading ability turn out to depend on its linguistic components.

Language, Speech, and Print

The linguistic nature of reading implies there should be something shared between print and speech processes. Indeed, there is a good deal in common between written and oral language. This assumption of commonality can be sloganized into such claims as "Print is speech writ down" or

"Reading is listening plus decoding."

On the other hand, it is possible to be impressed by differences between spoken and written language. Consider the following list of differences:

1. In speech, prosodic (e.g, intonation) and paralinguistic (e.g., gesture) features help signal meaning. Such features are absent in print.

2. In print, memory demands are reduced because the text can be reinspected. Speech makes higher memory demands because its signal is transient.

3. Print conventions mark word boundaries with spaces. In speech, word boundaries are not always marked by silence.

4. Speech is typically part of a social interaction. Reading ordinarily is an individual activity.

5. The content of speech is seldom arbitrary because of its social component. There is a shared message context between speaker and hearer. The content of print is often arbitrary from the perspective of a child reading in a classroom.

6. The content of speech is modulated by the participants. The content of print is fixed by the writer.

7. In learning to read, a learner must master a conventional coding system which maps a writing system onto a language system (a phonemic system in the case of alphabetic and syllabary writing systems, a semantic system in the case of ideographic systems). Speech is a natural coding system easily mastered by any child within a speech community.

This list can be extended to include some other features. For example, there typically are syntactic differences between speech and print. Written language uses more nominalizations (Chafe, 1982) and more subordinate clauses (O'Donnell, 1974). However, such differences may derive from one of the seven differences listed above or from some other source. For example, syntactic differences may derive from the memory feature (2) in conjunction with social context features (5 & 6) that emerge in typical purposes of writing compared with speech. However, whether derivative or fundamental, the fact that syntactic patterns of print diverge from syntactic patterns in speech has serious implications for reading. This is especially true for children in the early stages of reading, who may encounter unfamiliar syntactic patterns.

A generalization may be possible concerning these features that distinguish print from speech. There seem to be two fundamental features, design features of print and speech, from which others seem to derive:

The physical design of the signal. The spatial-visual design of print contrasts with the temporal-auditory design of speech. Such differences as 1, 2, 3, and probably 7 derive from this feature.

The social design of the message context. There is a fundamental contrast between the socially interactive and pragmatically functional context of speech and the asocial, one-way communication of print. Differences 4, 5, and 6 are derived from this feature.

Thus there are both commonalities and differences between spoken and written language. The central commonality is their dependence on linguistic structures and processes. Semantic, syntactic, and phonological structures are the linguistic core of language that is used in both print and speech. In addition, pragmatic, real-world knowledge is necessary in both speech and print. Differences concern the relative importance of particular structures and kinds of knowledge that derive from physical and social design features.

Implications for reading ability

If we consider the child learning to read, there are two obvious obstacles deriving from differences between speech and print: the *decoding* obstacle and the *decontextualization* obstacle.

The fact that print requires the learning of a code is fundamental, especially for orthographic writing systems. There are strong opinions about the implications of this fact—how and even whether decoding should be taught, for example. Some have argued that because reading is ultimately about comprehension, children should read only when comprehension is possible. Kenneth Goodman (1967) and Frank Smith (1973) are prominent proponents of this view.

By this view, the code, i.e., the decoding rules mapping letters to phonemes, is not to be taught. Reading is to occur only with texts, and the child's linguistic talents are supposed to allow him or her to use context to handle any problem of word recognition. This issue is too complex to be discussed here (see Chapters 10 and 11). However, one thing can be made clear: The child somehow has to discover the coding principles of his writing system. It is either impossible or difficult to become a skilled reader without mastering the code system. Young children with difficulty in reading almost invariably have imperfect knowledge of the coding system that maps written symbols to speech.

The second major obstacle is that the reader has to learn to deal with the decontextualized nature of print. He comes to school with a fairly rich knowledge of language and considerable experience in using it. However, this experience has been with contextualized language, speech that occurs in social context. He has seldom had to deal with meanings that are solely "in sentences." Meaning has been a joint venture, with context providing as much information concerning message interpretations as speech itself. More often than not, the things referred to in a conversation are physically present in the environment. When they are not, they are present in the world shared mentally by the participants in the conversation. Even speech the child hears on television is embedded in rich social and narrative contexts. It is quite possible that such contexts, since they overdetermine the interpretations of spoken messages, allow syntactic processing to be relatively unused and underdeveloped.

The child's encounters with print change all that. The meanings are less determined by context and more dependent upon sentences. The written sentence, unlike the spoken sentence, carries the meaning intrinsically. Syntactic processes must be used and they may not be sufficient. Also, inferences which might be supported by context in speech will be less available in print.

All of this is, of course, quite a simplification. But there is a central point concerning learning to read. Because of differences between print and speech, the child learning to read faces at least two obstacles: mastering a conventionalized coding system and adapting, eventually, to decontextualized language. In the course of these adjustments, reading will become increasingly a generalized linguistic activity. Differences between print and speech will be reduced. It is even likely that mastering reading leads to some changes in the way speech gets used.

Reading Ability: A Psycholinguistic Perspective

With the assumption that reading includes important linguistic processes comes the implication that reading ability can be understood in terms of linguistic processes. This implication constitutes a psycholinguistic perspective on reading and reading ability.

"Psycholinguistic" must be understood in a very careful sense. To call reading a psycholinguistic process is to assume that linguistic structures are important in the structure of written language and that language processes are important in the processing of print. It does not mean that reading is a "psycholinguistic guessing game," the description of reading suggested by Goodman (1967).

In fact, the metaphor of the "psycholinguistic guessing game" is an obstacle to understanding reading. The basic idea of the guessing game is that there are many cues that a reader uses to get at the meaning of text. Indeed there are syntactic cues, semantic cues, and graphemic cues. All these can be used, and are used, in reading. However, these "cues" depend on the reader's knowledge of linguistic structures and of concepts in intricate interaction with his knowledge of word structure—not just the arrangement of letters in the spelling of a word but also phonology and morphology. The reader can know, for example, that *nation* and *national* have different phonetic shapes (pronunciation) that are related to morphophonemic rules.

The intricacies of these linguistic systems, and especially how they are used in processes of reading, require investigation. The processes are complex and insufficiently understood. However, the psycholinguistic processes add up to something more than a guessing game. The skilled reader has quite a bit of certain knowledge at his disposal. The orthographic system provides a constrained set of possibilities for any given string of letters. And the coding principles provide very narrow choices for any orthographic string. Of course, *lead* may map onto "led" or "leed" out of context. But it can't map onto "window" or "deer." The skilled reader has adequate knowledge to identify most words without context and adequate knowledge to identify almost all words with very minimal context. For the skilled reader, reading is psycholinguistic but it is no guessing game.

Reading Ability: A Cognitive-Developmental Perspective

The psycholinguistic perspective of reading is appropriate because of the importance of language processes in reading. Another perspective overlaps with it. Reading is a process that includes many general cognitive components, especially with respect to comprehension. The knowledge that a reader has plays a large role in the understanding of a text. This knowledge is both specific to certain content (e.g., cooking and politics) and general to forms of thought. Both kinds of knowledge undergo development. The child reading at age 7 and the child reading at age 12 differ in the knowledge they bring to the reading task. Assuming each has adequate abilities at lexical access (word recognition) and given the same text, their comprehension of the text may be different.

Nevertheless, it is too ambitious to try to understand all of cognitive development. We will be concerned primarily with those aspects of reading that are most naturally associated with reading and language. If general cognitive development sets limits on reading ability, so be it. We want to understand reading ability within these limits. This means we want to understand especially how processes of lexical access, knowledge, and comprehension interact.

Reading Ability and Verbal Efficiency

Consideration of reading as cognitive and linguistic processes in interaction leads to the formulation of verbal efficiency. At least some of the processes in reading are executed within limitations of processing resources. Thus one key to the development of reading skill is the extent to which the limited-resources problem can be overcome.

With respect to learning how to read, mastery of the code is essential. As development of skill increases and as text demands increase, these coding processes do not become less important. They merely change in the nature of their impact. These coding processes must be executed fluently, with little effort. The increased textual demands mean the reader must take more and more use of higher level comprehension processes. This can be done only without expenditures of resources at low-level coding skills. This is the central claim of verbal efficiency theory (Chapters 6 and 7). Readers of low ability have inefficient—slow and effortful—coding as the major obstacle to reading achievement.

Reading Ability: Individual Differences

The range of individual differences in reading ability is enormous. Consider for a moment speed of reading as an index of ability. The average college reader reads at about 250 words per minute with some comprehension. However, some people read at 400 words per minute, while many others plod along at 150 words per minute. Of course, reading rates depend on reading materials and reading purposes, but these factors can be disregarded for now. The fact is that even among such a select population as college students, reading rates differ widely among individuals. One defining trait of a skilled reader is a reading rate that is at least average for the comparison population. A college student who reads at more than 250 words per minute over a wide range of texts is a skilled reader.

Of course speed is not enough. Some level of comprehension must be included. Our intuitive concept of reading skill includes reading fast and reading with good comprehension. One way to combine these two characteristics is to follow the lead of Jackson and McClelland (1975), who defined "effective reading speed" as the product of comprehension × words per minute. Thus two people could be equivalent in effective reading speed, the first twice as fast as the second and the second twice as good at answering comprehension questions.

Eventually, a theory of reading skill may be able to account for speed and comprehension separately. It is likely, for example, that the reader who puts speed as a priority is not only comprehending less but is engaged in slightly different processes (Just, Carpenter, & Masson, 1982). For now, both comprehension and speed must be considered part of reading skill. To an extent, reading rate may be traded off against comprehension. An individual can slow down or speed up by altering his criterion for comprehension. To this extent, then, it is quite sensible to define reading skill as including either high comprehension or high rate, with the other of these components above some minimum. Here we will seldom be able to refer to specific measures of speed and comprehension. It will, in fact, be common to have only measures of comprehension. In such cases we need only assume that a reader was not particularly slow, a condition often assured by time-limited standard assessments.

There is another issue of definition. Obviously a college-age reader who is highly skilled need not have just the same skills as a 10-year-old reader of high skill. There are qualitative differences in skilled reading that depend on the development of reading ability. Thus, a college reader may not simply do the same things that a 10-year-old does except faster. He may do them differently. There is no simple solution to this problem, because the solution depends on a deeper understanding of the very qualitative differences that might be involved. Instead, we must resort to a normative concept: Reading skill is relative to the age group of the reader.

Thus we have a definition of reading skill. A skilled reader is one who, relative to a given age group, shows comprehension and reading rates that are at least average. The less skilled reader, accordingly, is one who is below average in comprehension and/or reading rate. Although it is sometimes difficult to be sure whether the subjects in any particular study fit this definition, or whether they are fully comparable to the subjects in some other study, the definition is at least the ideal for any theoretical claims. It identifies the groups we refer to by the terms "skilled" and "less skilled" readers or the equivalents "high-ability" and "low-ability" readers.

SUMMARY

This introductory chapter has set forth the issues of reading and reading ability that are treated more fully in subsequent chapters. Reading ability is to be understood in terms of the cognitive processes of reading. Lexical access (word recognition) processes are those that identify words, and comprehension processes are those that build meaning representations of the text. Individual differences in reading ability can be understood only in reference to these processes.

Another important perspective on reading ability is that it reflects essential language processes. Linguistic processes, defined as the manipulation and representation of linguistic structures, are central to reading and therefore to our understanding of reading ability. In that sense the approach to reading developed in this book is a psycholinguistic approach. However, the linguistic components of reading are intricate. For example, there are both important similarities and differences between written and spoken language that derive from their basic design features. The beginning reader has to master a conventionalized coding system and learn to deal with decontextualized language, two prominent departures of print from speech.

Another possible perspective on reading ability is in terms of cognitive development. Cognitive capabilities set some limits on reading achievement. The main objective, however, is to understand the linguistic and cognitive components of reading processes that occur within these general constraints. An important assumption is that these processes take place partly within a limited-resource processing system. This leads to the verbal efficiency theory of reading ability.

Finally, the definition of reading ability comes from considering both speed and comprehension. Generally, a skilled reader is one who is above average in comprehension with at least average reading speed relative to a given age group. As a practical matter, measures of comprehension are sufficient.

CHAPTER TWO

Lexical Processes in Skilled Reading

A prominent feature of reading is that it involves encounters with words. Accordingly, understanding how these encounters work in skilled reading is a starting point for understanding reading ability. In what follows, the lexical processes of skilled reading are discussed, always in the perspective of how they work in texts.

Lexical Processes and Text Processes

In many cases, reading a single word is a significant accomplishment. An 8-year-old child who, upon seeing *contestant* can unflinchingly produce /kun TEST ənt/ has performed an impressive reading achievement. Why such an achievement has been so scorned by some (cf., Goodman, 1970; Smith, 1973) is a puzzle. In fact, the ability to decode words consistently—not by chance—is the essential reading process.

However, reading single words, no matter how impressive because of the novelty of the words (*rogation; pusilanimous*) or the inexperience of the reader is not what we really think of as reading. In a way, it is nothing more than practice for real reading—something like hitting balls over the fence in batting practice. The skill is there, but will it function in the game? The "game" in reading is comprehending a text. Not merely pronouncing the words, but understanding the text, especially when reading silently. A text is any series of coherently arranged sentences. Some texts tell stories, some explain how to assemble a kite, and others pretend to explain complex phenomena such as reading.

Thus our discussion focuses on a person who is reading a text of some sort. We want to know what processes bring comprehension of the text in a reader skilled at this task.

In accounting for the ability to read a text, there are two general classes of processes to consider. *Lexical processes* identify a word, including its constituent letters, and activate its semantic properties. The semantic encoding that results is the link to the other class, *comprehension processes*, which themselves include a number of different processes, some of them not so different from lexical processes. The two systems, the lexical and the comprehension, work together. Indeed they are linked by *semantic encoding*, the "comprehension" of contextually appropriate word meanings.

LEXICAL PROCESSES IN SKILLED READING OF TEXT

There is a general impression that we read words in bunches, skipping over many words. We read selectively, by this account. This impression does not reflect reality. The fact is that when we read the eyes come to rest on *(fixate)* most of the words of the text. Not many words are skipped.

There have been many studies of eye fixations and they reveal some important facts, some counterintuitive, about what the eyes are doing. The two most important may be these: (1) In normal reading, most words are fixated. (2) During a fixation, only limited information can be obtained from the visual periphery. Beyond five or six character spaces to the right of the fixation, letters are not perceived. There is a third fact worth noting: (3) Little information concerning words or letters is obtained during the eye's movement from one fixation to another (a *saccade*). Fact number 1 seems to be to be determined by facts 2 and 3. If information is obtained only during fixations, and then only within a few character spaces right of the fixation, then successful reading depends on fixating many words, not just a few.

To illustrate some facts about eye fixations, Figure 2-1 shows the eye fixations of a college reader studied by Carpenter and Just (1981; also Just & Carpenter, 1980). These are not actually individual fixations, but rather "gaze durations" that sum over individual fixations within a word. The fixations are numbered in order in parentheses. The numbers above each word are the durations (in milliseconds) for each fixation. It should be noted that the procedure used by Carpenter and Just to measure fixations is not universally used in eye movement research and that the measurement issue is important for theories of reading based on eye fixations (Kliegl, Olson, & Davidson, 1982). However, the conclusion that most words are fixated does not seem to depend on this measurement issue.[1]

There is more of interest than the frequency of fixation. Carpenter and Just (1981) estimated that about 80% of the text's content words (nouns, adjectives, verbs, adverbs) were fixated. However, the duration of a fixation is variable. Long words are fixated for a longer time than short words,

```
        (4)   (11)
        286   466

        (1)   (2)   (3)   (5)      (6)      (7)      (8)      (9)
        166   200   167   299      217      268      317      399
```

Radioiostopes have long been valuable tools in scientific and medical

```
                            (5)
                            183
     (10)              (1)   (2)   (3)   (4)        (6)   (8)  (7)    (9)
     463               317   250   367   416        333   183  450    650
```

research. Now, however, four non-radioactive isotopes are being produced.

```
                 (4)                        (8)
                 366                        183
     (1)         (2)   (3)      (5)      (6)        (7)   (9)   (10) (11)
     250         200   367      400      216        233   317    283  100
```

They are called "icons"—four isotopes of carbon, oxygen, nitrogen and

```
(12) (13)
683  150
```

sulfur.

Figure 2-1. Eye-fixation pattern of a college student reading a technical passage. The numbers 1,2,3 . . . indicate the sequence of fixations. Thus, number 4 is a regressive fixation, i.e., the result of the reader's gaze going back to the first word. The numbers below each fixation number represent the duration of the fixation. For example, fixation number 7 was on *tools* for 268 msec. The subject read with the expectation of a true-false text. Adapted from Carpenter and Just (1981).

and infrequent words are fixated longer than frequent words. Also, the last word of a sentence receives a longer fixation than other words. In Figure 2-1, *research* and *produced,* each the final word of its sentence was fixated about half a second, much longer than the average fixation. Carpenter and Just (1981) refer to this as "sentence wrap up" time. The wrap-up time appears to capture extra processing in which the reader is assembling sentence parts or adding interpretations to parts of the sentence that were incompletely or incorrectly interpreted. In fact, the Carpenter and Just model assumes that each word is semantically encoded as much as possible when it is initially encountered. There is no "buffering" for later interpretation.

Visual and lexical access factors

The fact that linguistic variables such as word length and word frequency have an effect on fixation times is important for a theory of reading. It implicates specific properties of words that are part of the reader's memory representation for words. It also encourages the assumption that linguistically based lexical factors are critical in reading ability. On the other hand, it is important to realize that these lexical processes are also visual processes. For example, the fact that the eye spends more time (makes more discrete fixations) on a long word than a short word is partly due to the requirements of visual sampling. The longer a word is in letters, the more sampling may be required. Kliegl et al. (1982) found that the number of fixations (hence the "gaze duration") was at least as related to the number of letters in a word as to the number of syllables. They point out that the number of fixations a word receives is partly a function of whether an initial fixation was "inconveniently" placed near the beginning or end of the word. After an "inconveniently" placed fixation an additional fixation is more likely to occur. The study of eye fixations during reading is providing a bit of information about the cognitive, linguistic, and perceptual demands of the reading task. All things considered, it seems clear that the number of discrete fixations on a word, and hence the total duration of fixations, reflects linguistic and cognitive demands of lexical access. That is, some of the actual fixation time is going to cognitive processing (Rayner & McConkie, 1976; Rayner, 1977). On the other hand, the essentially visual demands of the reading task and their effect on directing the eye to its next fixation should not be underestimated.

Lexical Processes in Rapid Reading

The purpose of the reader has an effect on these eye fixations. The fixations of Figure 2-1 are for a reader who expected a true-false test following reading. When readers are "skimming" the text without such an expectation, they show fewer and shorter fixations (Just, Carpenter, & Masson, 1982). However, they show a similar sensitivity to word-level variables such as length and frequency.

In fact, eye movement studies of speeded reading show little support for the widespread belief that the skilled reader can process texts at very high rates while maintaining acceptable levels of comprehension. Just et al. (1982) report eye movement studies of readers who have been trained in a speed-reading course. They had had about 50 hours of practice at rapid reading. These trained speed readers were compared both with normal readers and with readers instructed to read rapidly using their own techniques (skimmers). Both the skimmers and the speed readers read at 600–700 words per minute, compared with the normal readers' 250 words per minute. Note that the speed-reading and skimming rates are rather modest when compared with anecdotal and even scientific reports of very high rates of 5,000 words per minute and more. It appears that to achieve such very high rates, readers are in fact skipping enormous chunks of text as they follow odd patterns down and up the page. The fastest reader reported in the literature appears to be one described by Thomas (1962) who read at a rate of 10,000 words per minute. According to Thomas' report (described in Just et al., 1982) this reader made an average of only six fixations per page, going down the left side and up the right side and skipping the bottom one-third of the page altogether! Studies by McLaughlin (1969) and Taylor (1962) on somewhat slower speed readers tell a similar story. For example, McLaughlin's (1969) 3,500-word-per-minute reader averaged fourteen fixations per page.

Compared with such rates, the speed readers of the Just et al. (1982) studies were plodders. Still, they read at two to three times the normal rate. How did they read? There were two important departures from normal. The speed readers fixated fewer words, only about half the words fixated by normal readers, and their average fixation duration was about two-thirds as long as that of normal readers, about 230–240 milliseconds per fixation compared with about 330 milliseconds per fixation for normal readers.

There were some important similarities as well. Speed readers, like normal readers, fixated content words more frequently than function words—about 50% of the content words and about 25% of the function words. Normal readers had corresponding figures of 77% and 42%, respectively. (All the data are based on averages of two very different passages, one a *Reader's Digest* story and the other a *Scientific American* article. The differences in fixation data between the two passages were very small.)

Thus the speed readers made fewer and shorter fixations but distributed them similarly with respect to content and function words. Are their lexical and comprehension processes the same as those of normal readers, except faster? Apparently so, with some qualification. We can summarize the answer suggested by Just et al.'s data by referring to *lexical difficulty, selectivity, adaptability,* and *comprehension.*

Lexical Difficulty. As with normal readers, the fixation durations of speed readers were affected by the length and frequency of a word. But speed readers were less affected by these factors. They may in fact, as Just et al. suggest, have had a "processing deadline" for words so that the lexical factors of length and frequency would exert an effect, but a limited one.

Selectivity. Also like normal readers, speed readers were not selective about fixating "important" words in the text. This, too, counters common conceptions. The eye is not directed to words that will turn out to be more important in the text. It has to discover what is important, at least at the word level.

Adaptability. There was an increase in the fixation rate in sections of higher text difficulty for both normal and speed readers. However, speed readers did not change much in a difficult section. They increased their sampling rate on a difficult "page" (a screen, actually) but they did not modify durations nor abandon their strategy of uniform sampling. (This contrasts with the skimmers, who sometimes reverted to more normal patterns.)

Comprehension. Speed readers did well enough at answering questions on high-level information (they were better than skimmers at getting gist). However, they did less well at responding to questions requiring detail. Generally, they could not answer a question unless they had fixated within three character spaces of the answer during reading. Since they fixated less material in the text, their comprehension had to be less.

Summary. These eye movement studies should take much of the mystery out of speed reading. Comprehension that depends upon explicit text information cannot be maintained without fixating on many text words. Since reduction of fixation frequency is a major contributor to the faster rates, it follows that comprehension must undergo some reduction. On the other hand, it is equally clear that speed readers can learn to control fixation durations and frequency in such a way to sample enough information to obtain the gist of a text.

The Perceptual Span

As we have seen, a normal reader fixates up to 80% of the content words in a text. Furthermore, a reader who samples less of the text also comprehends less of it, as the studies of speed reading demonstrate. Why is so much text sampling necessary?

Part of the reason is that the span is relatively narrow. The perceptual span is the spatial extent from the central fixation where some information is obtained. The narrowness of the perceptual span comes as another of those surprises to our intuitions, in this case our feeling that we can read words at the far side of the page (the visual periphery) when we are focusing on the center of the page. In fact, this span appears to be only a few letters, perhaps as few as three (Rayner, 1975). This conclusion comes from studies using visual displays that could be changed in response to a subject's fixation on any given location. Thus the perceptual span can be inferred by altering the text, e.g., changing a letter four character spaces away from the reader's fixation. If the alteration affects the reader by lengthening the duration of his present fixation, or if his next fixation indicates an effect of the change in some way, then his perceptual span is at least four character spaces long. Rayner (1975) found that a letter alteration that changed one word into another had little effect when it was beyond three spaces to the right. However, Rayner (1975) concluded that some information concerning word shape and specific letters was obtained out to twelve spaces from the fixation. Word-length information is also noticed at the distance (McConkie & Rayner, 1973).

As more evidence on perceptual spans has been obtained, only very minor variations in the span estimations appear. It seems certain that specific letter and word information is obtained from only a relatively narrow range beyond the fixation point. Furthermore, comparison of children with adults shows that this span does not change much with increasing reading skill (McConkie, 1982). However, this does not mean that useful information is completely restricted to the word being fixated. There is evidence that some information in the next word beyond the fixation can be used. Rayner, McConkie, and Zola (1980), using the text alteration procedure described above, found that the first three letters of a word presented parafoveally facilitated recognition of the word on the next fixation. While the word in the parafoveal region could not be identified, the reader was obtaining partial information from its initial letters that could be used to identify it when it was actually fixated. Furthermore, if the parafoveally presented word merely had letters visually similar to the word presented when the eyes moved to its location, there was also facilitation.

Thus, at the edge of the narrow foveal span where words can be identified, the reader is able to use partial information from the next word. In a sense, the reader may be processing more than one word on a fixation, although only the fixated word is "accessed." We can think of this as a preactivation of the letters contained in a word that reduces the amount of processing needed to identify it (Rayner, Well, Pollatsek, & Bertera, 1982). At the same time, it seems correct to say that reading, as lexical access, is confined to the word being fixated.

Summary

This section has demonstrated the importance of *lexical processes* in reading. The eye movement research is perhaps uniquely able to inform us about the significance of individual words for reading texts. (See Rayner [1978] for a thorough review of eye movement research.) Normally, most words are read and words beyond the one being read are not processed sufficiently for their meanings to be encoded. Thus lexical access, the process by which a word is recognized, is the central recurring part of reading.

LEXICAL ACCESS

In this section, we consider the processes used to access a word in permanent memory. The starting point for lexical access in the case of reading is a visually available input. Languages, of course, differ in how they graphically represent the units of language. Some are logographic, representing a word-concept with a single sign, and others are syllabic, assigning a meaningless spoken unit to each written symbol. However, most modern systems are, like English, alphabetic systems. In an alphabetic system, the graphic signs correspond, however variably, to single, abstract speech sound units, or phonemes. In discussing lexical access in this section, we are referring to access in an alphabetic orthography. However, because the level of description will not be highly detailed, much of it will apply to lexical access in other systems.

First, a definition of *lexical access*. Lexical access and word identification will be used interchangeably. Or rather "lexical access" will be taken to include word identification. These two processes ordinarily are distinguished, but this distinction is largely a matter of the experimental paradigms which give meaning to the terms. Thus word identification has meant the recognition of a letter string as a particular word, while lexical access has usually meant the recognition that some letter string is a real word rather than not a word. It seems clear that it is word *identification* that is relevant for reading. The reader has to identify, somehow, the words he encounters, not merely register their occurrence. The reason for using the phrase "lexical access" rather than "word identification" is that it is more theoretically neutral, once it is acknowledged that mere access (dictionary registration) is not enough for reading. So here is the working definition: *lexical access* is the process of finding a written word in long-term memory. It initiates the critical processes of *semantic encoding*, i.e., attaching a contextually relevant meaning for the word to the ongoing text processing. It initiates also the process of phonetic activation that may also play a critical role in reading.

The processes of lexical access and word identification have been very widely investigated and a number of contested issues have developed. These include especially the role of speech recoding prior to access and whether there are whole-word processes independent of individual letter recognition. For the most part, the discussion of lexical access will ignore these issues, developing instead a perspective that makes them less critical for the theory of reading ability. (For a review of some of the issues of word recognition, see Baron, 1978). The perspective needed is to consider lexical access the result of interactive processes.

Interactive Processes in Lexical Access

Interactive processing occurs when information at different levels is combined to jointly determine some outcome. In the present case, the outcome is access to a word's location in permanent memory. An interactive model of lexical access can be of the *strong* type (strongly interactive) or the *weak* type (weakly interactive). *Strong* and *weak*, of course, are not properties of the model's power but rather of its processing assumptions. A *strongly interactive* model assumes that processes are mutually influenced at all levels. Lower level processes affect and are affected by higher level processes. A weakly interactive model assumes only that multiple sources of information affect a final outcome. These sources either do not actually affect each other or, if they do, they do so only in one direction, from lower level to higher level. These two types of weakly interactive models are represented, respectively, by the logogen model of Morton (1969) and the cascade model of McClelland (1979). It is also the kind of model described by Perfetti and Roth (1981) specifically to account for individual differences in reading ability.

The weakly interactive models will in fact do a rather good job of accounting for the facts of word recognition. So too might some stage models, such as that of Massaro (1978), since despite their names, stage models often seem to contain weakly interactive assumptions. The point here is not to assess which models account for most basic facts of lexical access nor even to compare the models. However, it seems likely that the facts of recognition and access are best handled by some model that assumes that processes interact. The model described by Rumelhart and Mc-Clelland (1981; 1982; McClelland & Rumelhart, 1981) serves this requirement very well. It is a model of the strongly interactive type and perhaps a more powerful model than is actually needed to account just for lexical access.

Figure 2-2 shows the essential features of an interactive model, based on the proposals of Rumelhart and McClelland. Its key assumptions are (1) that each level of information—grapheme, phoneme, word—is separately represented in memory and (2) that information passes from one level to the other in *both* directions. (This is what makes it *strongly* interactive.)

An interactive model of this type contrasts with a strictly "bottom-up" model. In bottom-up processing, information goes only from the lower levels to the higher levels and not the reverse. Thus, detection of lines and angles leads to identification of letters, which in turn leads to iden-

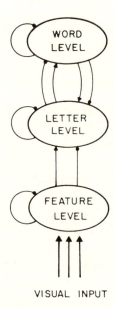

VISUAL INPUT

Figure 2-2. Levels of information of the interactive word-perception model of Rumelhart and McClelland (1981). Activation travels up to the word level from the letter level and down to the letter level from the word level. (Inhibition also occurs within a level—e.g., activation of *H* inhibits other letters—and between letters—e.g., activation of the word *rest* inhibits g and all other letters not linked to the word.) A phoneme level, omitted here for simplification, may be necessary for some lexical processes (See Chapter 4).

tification of words. Such processes are, of course, essential to reading because they allow the reader to identify what is actually on the page—to read instead of hallucinate.

The interactive model, of course, includes bottom-up processes. It differs, as can be seen in Figure 2-2, in allowing *activation* to travel from words to letters, as well as from letters to words. The links between levels are both excitatory and inhibitory. Thus as the lines and angles (visual features) are detected, activation travels up to letters that are consistent with these features. For example, very early in processing, the letters *H, E, F* would all be activated if the feature level passed on the information that there was a vertical line bisected by a rightward extending horizontal line ($\vdash$). Activation of other letters would be low. As the feature level detected a second vertical line in a certain location, the activation of *H* would greatly increase and the *net* level for *E* and *F* would decrease.

The same activation principle results in the activation of candidate words, i.e., words consistent with all information available at any precise point in the process. In the example given, all words having *H, F,* or *E* would receive some activation that pushes them beyond their resting level. One can assume, as Rumelhart and McClelland do, that a word's resting activation level will be a function of its familiarity, or frequency of encounter. It will take more activation to access *martyr* than *martha* because its resting activation is lower for most readers. It is possible that what is responsible for the resting activation level is the *recency* of an encounter rather than its *frequency*. Of course, as more feature information gets detected, the activation of letter candidates and word candidates changes. One word finally wins out. The operative principle seems to be that many are activated, few are accessed.

This much is bottom-up processing. What makes the processing interactive is that activation travels down from words as well as up to words. Suppose, to follow the earlier example, *H, F,* and *E* are all highly activated, and at the word level, THE, OFF, and DEN are among those with some activation because of having a highly activated letter in the second position. Of course the letter level contains *T, H, E, O, F, P, N,* the letters found in the activated words. These letters are increased in activation even though none of them, except those consistent with ⊢ , have been "seen." Because of this "top-down" (word to letter) activation, less feature information is now needed to "see" a *T* or *P* or any of these letters. And since this indirect activation of these letters is added to that being initiated at the feature level, the final perception of the word has been speeded up. It is faster than if each letter had to be completely identified before the words could be identified.

Context Effects

This interaction of letter level and word level is the part of an interactive system that has been worked out in detail (Rumelhart & McClelland, 1981; 1982; McClelland & Rumelhart, 1981). However, there is another level to consider that is at least as important for skilled reading. Suppose the word level can also receive activation from context. For this, we can imagine a text memory buffer, perhaps working memory, that, as it processes the text, activates the words in permanent memory as they are encountered. Activation spreads to semantically related words. This means that a given word's activation might well be above resting level prior to any information whatsoever from the word itself. In fact, at least one other word-recognition model makes this sort of assumption (Morton, 1969) and a mechanism of spreading semantic activation is a central processing assumption for which there is much evidence (Collins & Loftus, 1975). This kind of semantic influence was also a part of an earlier interactive model of Rumelhart (1977), and its application to the Rumelhart and Mc-Clelland model is consistent with the assumptions of the theory.

A semantic influence of this sort means that identification of words will be facilitated in context. Such context effects are well documented (e.g.,

Morton, 1964; Perfetti, Goldman, & Hogaboam, 1979; Schuberth & Eimas, 1977; Tulving & Gold, 1963; West & Stanovich, 1978). The application of context can be only weakly interactive. It simply lowers the threshold for perceiving a word. In the strongly interactive model it would operate to affect activation levels of letters that are part of activated word candidates as well as the words themselves.

In short, whether by weakly or strongly interactive assumptions, this top-down semantic effect makes a difference for lexical access in reading. For example, the word ~~window~~, which appears here as a blur to represent a stage of feature processing that is incomplete, is hard to recognize. But if it occurs in sentence (1) it is a bit easier to recognize:

(1) *There were several repair jobs to be done.*
       The first was to fix the ~~window~~.

However, even this context is less helpful than (2):

(2) *The room was warm and stuffy, so they opened the*

As the text increasingly constrains the choice of words—(2) is more constraining than (1)—the top-down activation is spread less thin. In (1) semantic activation is spread among many words—*window, chair, bicycle, hot water heater,* etc.—that refer to repairable objects. In (2) semantic activation is more concentrated, and *window* receives most of the activation; *door* perhaps receives some also.

Thus, by this account, the access of a word in memory is the result of semantic processes as well as processes that identify letters strictly on the basis of feature information. Semantic information may in fact compensate for impoverished feature information, as the *window* example above demonstrates. For example, Perfetti and Roth (1981) report experiments based on just this principle, in which children read words that were visually degraded by randomly deleting features. (Since these were computer-printed words, the features were essentially dots.) The degree of stimulus degradation could be varied by the percentage of dots that were deleted. Figure 2-3 shows an example of three levels of degrading. At the highest level of degrading, words were very slowly identified and often not identified correctly at all. When the word appeared in context, it became more identifiable, both in accuracy and speed of identification. The contexts were arranged to vary the constraint applied to the final word, which was always the word to be identified. Figure 2-4 shows the result that speed of identification rose with increases in the degree of constraint imposed by the context.

The interaction of contextual constraint and degrading suggests that high semantic activation can compensate for insufficient feature and letter ac-

0 %	pepper	window
21 %	pepper	window
42 %	pepper	window

Figure 2-3. Three levels of stimulus degrading used in word-identification experiments of Perfetti and Roth (1981). Identification times increased with degrading.

tivation. It makes another important suggestion as well. The fact that there was a statistical interaction of constraint and degrading can be taken to demonstrate that the contribution of semantic activation *increases* as lower level feature and letter activation *decreases*.[2] Theoretically, this is important because it is consistent with an assumption that the interactive system is asymmetrical. The asymmetry is that the lower level processes in skilled reading are *autonomous*. That is, they can make a contribution to word identification that is independent of context. To put it another way, the asymmetry in the interactive system means that low level in-

Figure 2-4. Identification of words related to degree of contextual constraint—high, medium, low, and words in isolation—and to degree of degradation.

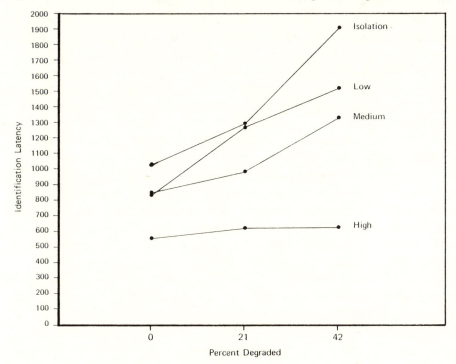

formation is *sufficient* for identification whereas semantic information is not sufficient. Free associating is not the same as reading.

To summarize, an interactive model of lexical access will account for semantic activation effects as well as the basic facts of isolated word recognition. The model of Rumelhart and McClelland has been worked out in detail only for isolated word recognition. Semantic activation, in which ongoing text processes affect lexical access, can be incorporated into such a model, although its processing features have not been worked out.

Context Mechanisms for Semantic Activation

If semantic activation actually assists lexical access, there are several questions to ask. The central question is how does semantic activation work? One possibility is that the reader actively anticipates possible word candidates during reading (a pernicious version of the psycholinguistic guessing game). This is possible in principle, but we are talking here about *skilled* reading. Skilled reading is, by definition, a very fluent process. If a reader fixates three or four words per second, around the normal rate, where is there time to guess? Moreover, if he is skilled at reading, why bother? Reading is much easier than guessing. The case may be different in, for example, reading in a foreign language that is incompletely mastered. There is plenty of time to guess in such cases and perhaps enough payoff for doing so. But for a skilled reader and a familiar language, there must be a mechanism other than conscious anticipation.

The alternative, of course, is automatic (unconscious) activation. The feature-detection and letter-activation processes that bring about lexical access from the bottom are of this sort. Automatic semantic activation could rise very rapidly in time to assist word identification. There are a number of models that describe such automatic activation during reading (Becker, 1976; Fischler & Bloom, 1979; Stanovich, 1981). These models are different in how they handle facilitative and inhibitory effects of context. Inhibitory effects refer to increases in processing time associated with a misleading context.

An example of an activation model that can account for context effects is the model of Posner and Snyder (1975), as applied to reading by Stanovich (1981). This model assumes two separate processing mechanisms that govern expectancy, one that is automatic, or attention-free, and one that is conscious, or attention-controlling. The two processes are assumed to have different time courses. The conscious attention process works more slowly because it has to direct a limited-capacity process to a new input. Or, in other words, it shifts the location of attention from one point to another within semantic memory (in the case of reading).

By contrast, the attention-free mechanism operates quickly and at no cost to the limited-capacity processor. The two-process theory accounts for inhibition effects by the conscious attention process and for facilitation effects by the automatic attention process. The implication that facilitative effects are more rapid than inhibitory effects gets some support from experiments by Stanovich (1981). Other research has been interpreted to support alternative models (Becker, 1980; Fischler & Bloom, 1979). Under what conditions inhibitory effects occur seems to be the dividing issue. However, it seems clear, at minimum, that there is some rapidly occurring activation function that can facilitate lexical access under the right conditions.

As an alternative to these concerns over inhibition, there is an assumption perhaps more suitable for reading, namely, that inhibition processes do not occur. (That is why the earlier discussion of context ignored inhibition.) Inhibition is a process that results from a really unexpected occurrence. If a woman is reading a book as she begins to cross the street, and a bus suddenly interrupts her reading, that is inhibition. A new signal must be processed, and attention is clearly shifted to a new location. But within the reading process itself, inhibition should be a rare event. Perhaps encountering a four-letter epithet in the middle of a *New York Times* column would be such a case. But for run-of-the-mill reading, this kind of attention-shifting inhibition should be rare.

By this alternative, conscious attention mechanisms are of very reduced significance.[3] What is significant is unconscious spreading semantic activation. Activation would spread among concepts in memory as the text meanings are encoded. The activation can be thought of as spreading along the nodes of a semantic memory network (Collins & Loftus, 1975). On such a system, we would want activation to occur for words other than simple associations. For example consider sentence (3):

    (3) It was really *hot*, so we *opened* the *window*.

It would be nice if *window* received some prior semantic activation (or "priming") from the sentence. *Opened* could initiate activation to things that are marked in a semantic network as [can be opened]. Perhaps *door*, *window*, *shutter*, *bottle*, *package*, and a number of other words get some activation from *open*. But if *window* is to get more than its share of activation, and the evidence suggests that is does, it must get an additional jolt from somewhere else. A likely source is *hot*. However, *hot* would not be directly linked to *window* in a semantic network. The spread of activation occurs along all links as the nodes are encountered, and *window* will receive some activation from intermediate links between its node and *hot.* Another source of activation lies in the previous sentence. The problem with assuming such distant context effects based on spreading activation is that the time course is probably wrong. Activation rises and falls rapidly. Whether there would be much effect after several words have intervened is doubtful.

There are two more contextual mechanisms that will help out. One is based on syntax. It is possible that at least some activation results from the initial encoding of a syntactic pattern. Parts of the pattern not yet encountered are activated. For example *the* would trigger a noun-phrase pattern. Any noun in the system gets a piece of the activation and so would any prenominal adjective. When an adjective is also encountered, e.g., *the old*, there continues to be activation for all nouns, supplemented by semantic activation from *old*. Words that cannot be nouns would receive inhibition (in the interactive model sense of inhibitory links, not an active inhibition). There is in fact some evidence for an automatic syntactic priming mechanism for definite articles (Irwin, Bock, & Stanovich, 1982). There are also strong effects of syntactic expectation on how a sentence is interpreted (Frazier & Rayner, 1982).

The third contextual mechanism is based on the assumption that the reader builds a mental model of the text during reading. That is, he attempts to comprehend the text by constructing a representation of the underlying text meaning. This mental model can be thought of as an abstract conceptual schema but with specific verbal information. For example, halfway through reading Little Red Riding Hood (for the first time) the reader's model includes information about a visit to a sick grandmother interrupted by an encounter with a wolf. This central thread has connections to many lower level information structures—for example, the mother-daughter episode at the beginning of the story—that recede as the text model gets updated. The context mechanism is that activation is available to elements in the text model. Activation rises and falls for particular elements as the story progresses. Thus the part of the model to which the most recent piece of information is attached is more active than other parts of the model. However, all elements of the model are active relative to the base level. Thus even though grandmother has not been mentioned for a few sentences, grandmother-related concepts are at least weakly activated. Lexical access for words linked to their concepts may be slightly facilitated. Alternatively, and more likely, access may be relatively unaffected but the semantic processes that assign contextually appropriate meanings to words may be significantly affected.

The force of this proposal is that context effects may occur both because of what has just been read and because of what is in the mind of the reader. The latter, in the right circumstances, will include a representation of the text.

What do context mechanisms affect?

Do context mechanisms work on lexical access or do they work on some process following lexical access? In conventional descriptions of the reading process, this seems to be an important question. Mitchell and Green (1978), for example, interpret the kind of evidence discussed in this section as reflecting postaccess processes (involved in saying the word, for example). They suggest that in normal reading, context plays little role in lexical access itself. To assess this issue, we can return to what the eye tells us about reading. Zola (1979; reported in McConkie & Zola, 1981) studied the eye movements of subjects reading paragraphs with highly predictable words. Some of the words were 85% predictable. What would the eye do, for example, on the word *popcorn* when it was preceded by *buttered* compared with when it was preceded by *adequate*? What it did *not* do is skip the word. The target word, e.g., *popcorn*, was fixated 96% regardless of how predictable it was. Contextual constraint did have some effect, however. When the word was highly predictable, the fixation was about 14 msec shorter. These context effects, however, do increase when the target word is short enough so that nonfixations are possible. Ehrlich and Rayner (1981) reported that target words are fixated less often when they are highly predictable. They also found a larger effect of predictability on fixation durations (over 30 msec). Of course it is possible that the savings of 14–30 milliseconds occur during some postaccess process. In particular, the contextually appropriate semantic encoding of the predictable word is made easier. But if this encoding is what is meant by a "postaccess" process, then it is certainly an important early occurring part of reading. If we accept a basic distinction between preaccess and postaccess then it may turn out that most semantic context effects are postaccess in ordinary reading.

As a matter of fact, the preaccess vs. postaccess distinction becomes less meaningful in a fully interactive model. Since activation is passing between multiple information sources *there is no clearly defined stage of lexical access.* There is simply some time after which activation is high enough to say that a word has been "identified." Meanwhile semantic activation, which has been going on during this time, may continue. So, roughly speaking, the semantic encoding of a word is part of its lexical access.

SUMMARY

This chapter has described some of the lexical processing in skilled reading. Skilled reading, defined as a combination of speed and comprehension, was examined by considering some basic studies of eye movements. Research indicates that most words are fixated during skilled reading, with content words fixated more than function words. Speed readers read faster by making fewer and shorter fixations. However, their comprehension of details suffers because only words that are fixated provide information about meaning. The perceptual span for specific word identification is only a few characters, although some shape-length information has a large span. These studies demonstrate the central importance of lexical access in reading.

Lexical access, defined as access to a word's location in memory, was described as an interactive process. Although weakly interactive models of access would serve well, the strongly interactive model of Rumelhart and McClelland was taken as a powerful and plausible description of lexical access during reading. Lower level and higher level information travel through a network along activation links in both bottom-up and top-down directions. Activation also is initiated by semantic processes. Thus access to a word in context, the usual case in reading, is facilitated compared with access in isolation. Experiments using trade-offs between context and word-level information demonstrate this important effect. Three possible activation processes are proposed: semantic activation from words in the sentence being read, activation from incomplete syntactic frames based on grammatical categories represented in the memory network, and a model of the text's meaning constructed by the reader. Evidence for these mechanisms is needed. The question of whether the context effects occur during "preaccess" or "postaccess" processes is less important in an interactive model.

NOTES

1. The issue is what counts as a fixation. Carpenter and Just (1981; Just & Carpenter, 1980) have used an aggregate measure (gaze duration) that sums all fixations beyond some minimum. It is more typical in eye movement research not to summate over fixations. According to Kliegl, Olson, and Davidson (1982), the procedure of Just and Carpenter (1980) produces misreadings of fixation time that cause problems for theoretical modeling. However, this controversy should not affect the conclusion that most words are fixated.
2. Interactions between contextual constraint and word-level factors are also found for word factors other than degrading. For example, Perfetti, Goldman, and Hogaboam (1979) found a word length × context interaction. Longer words were facilitated more by context than were shorter words. Stanovich (1981) found a similar result with the same measure, time to identify a word (word naming). Stanovich and West (1981) also report an interaction of context with degrading. However, this interaction involves an *inhibitory* effect (incongruent context) more than a facilitative effect. A reliable interaction for facilitation may be more de-

tectable with multiple levels of degrading and context, as were used in the Perfetti and Roth (1981) experiments but not the West and Stanovich (1978) and Stanovich (1981) experiments.

3. It also follows that the best way to assess context effects in ways that apply to normal reading is to compare different degrees of facilitative context with weak or no context. Experiments which have inhibitory instances mixed in with facilitative ones may not assess facilitation appropriately for reading.

CHAPTER THREE

Comprehension in Skilled Reading

This chapter considers the main features of comprehension in skilled reading. At the outset, an important point must be emphasized. The distinction between lexical access and comprehension is not one between two stages of processing. The skilled reader does not first access a word and then comprehend. Instead, in normal reading comprehension and lexical access are concurrent ongoing processes. The reader who accesses a word generates both an updated model of the text and an immediate bit of local processing. This assumption indeed is necessary for the proposal of the last chapter that a text model provides semantic activation.

There are two major components to comprehension: *local processing* and *text modeling*. Both of these components depend on the reader's knowledge of word meanings, knowledge of domains related to the text content, and processes of inferring. To discuss these aspects of comprehension, we will consider the following simple story excerpt. Each sentence is numbered for reference.

(1) *Joe and his infant daughter were waiting for the doctor to get back from lunch.*
(2) *The room was warm and stuffy, so they opened the window.*
(3) *It didn't help much.*
(4) *Why wasn't there an air conditioner?*
(5) *Strange for this part of Manhattan.*
(6) *Suddenly the door swung open.*

In this brief excerpt there are six sentences and 47 words. It is a very simple text that could be read quickly and easily by a skilled reader. What are the processes of comprehension as this text is read?

LOCAL PROCESSING

By local processing, we mean those processes that construct elementary meaning units from the text over a relatively brief period. These include the semantic encoding of words and the assembling of propositions. These processes, at least the second one, take place within a limited-capacity processing system, which can process only a limited amount of text at any one time.

Semantic Encoding

Encoding the meaning of a single word is of course a lexical process and might as well have been discussed in a chapter on lexical processes as in one on comprehension. However, the meaning of a word is encoded in a way that is appropriate for its context. This is comprehension.

Actually, what a reader must have at the outset is an entry for each word in semantic memory. Such a semantic memory may be conceived as a network that links word concepts to abstracted meaning features and to other word concepts. The exact form of this representation is not important, although there certainly are a number of different systems (Anderson, 1976; Lindsay & Norman, 1977; Smith, Shoben, & Rips, 1974). In fact, the two main representation systems, semantic features and semantic networks, are formally nearly equivalent although they appear different on the surface.

Figure 3-1 shows a piece of a semantic memory system that would be activated during the reading of sentence (2). Its representation is something of a mix of current systems, being most similar to that of Lindsay and Norman (1977) except that it represents meanings directly (to the right of each concept) in addition to links to other word concepts.[1]

The reader has to have some word knowledge of this sort, however one chooses to represent it. The representation must provide sufficient structure so that while *hot* and *cold* are linked, the reader can say something like "*hot* is the opposite of *cold*" and not "*hot* means *cold*." He must also have knowledge that includes what is usually called *presupposition*. For example, to understand the meaning of *opened* in sentence (2), the reader presupposes that the object to be opened was closed. Such knowledge can be represented in different ways, perhaps best as an inference system that operates on the semantic network. (These matters have not received much attention in semantic network theories.)

Context-appropriate meanings

In sentence (2), the word *room* has more than one meaning, only one of which is represented in the network of Figure 3-1. It can also mean,

Some Linked Concepts	Directly Activated Word Concepts		Characteristic Features

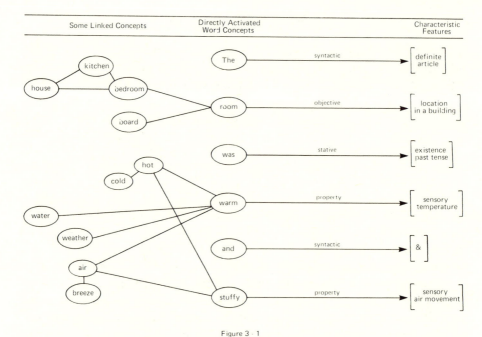

Figure 3 · 1

Figure 3-1. A piece of semantic memory network activated by *The room was warm and stuffy.* Word concepts are nodes in the memory network. They are closely linked with meaning features that "define" the concept. (This is a departure from network representations such as Lindsay & Norman's [1977], in which labeled links exist only among concept nodes.) Meaning features are connected by links that label the general class of each concept. These include *syntactic modulators, actions, states, properties, objectives, abstractions,* and *names.* To the left are links to other concepts that potentially receive activation.

roughly, *open space without restriction,* as in "The university has no more room for expansion." It also has common metaphorical extension, as in "room to grow." Such multiple meanings are very common in natural languages. Furthermore, the multiple meanings of a word are not always so closely related as in the room example. The word *bar,* for instance, has two meanings with little in common, as both lawyers and bartenders will attest.

How does the reader encode the word in its appropriate meaning? The intuitive answer is that context determines this process. When one is reading about an attorney's struggles with an ethics committee, *bar* is encoded as [lawyers' guild]. When one is reading about the best places to have a pina colada in Philadelphia, *bar* is encoded as [drinking place].

This is undoubtedly true, but when does context do its work? It would be convenient if there were always powerful semantic cues before the word is encountered. But what about when the word is inconsiderate enough to occur early in a sentence? *Room* in sentence (2) is such an example. Often, there is also context assistance from the previous sentence, but not always: *It was a beautiful day. The fair. . .* At just this point how will *fair* be encoded? Certainly the remaining context will force an appropriate encoding of *fair*. But at the time of lexical access, there are many possibilities that fit the context. *The fair weather has been with us for a week. The fair Gwendolyn would be mine by nightfall.*

The solution is in the activation model. As context accrues, semantic activation rises and falls. If context heavily biases one meaning over another, then that appropriate meaning will have an activation advantage. However, because activation spreads, even inappropriate meanings will receive some activation. So two things should happen: (1) All meanings should be activated, and (2) the appropriate meaning "wins out" strictly as a function of the rate of its activation and the rate of decay of activation for the inappropriate meaning.

The first of these predictions, counterintuitive as it is, seems to have been confirmed. Swinney (1979) had subjects perform a lexical decision task, i.e., to decide whether a letter string was a word or a nonword. The lexical decision was made visually just at the moment that an ambiguous word was heard through a headset. It is known that lexical decisions are affected by prior semantic activation. Thus *butter* preceded by *bread* is accessed more quickly than *butter* alone, and the lexical decision is faster, relative to a control (Meyer & Schvaneveldt, 1971). The question is whether such semantic activation would include both meanings of an ambiguous word or just its contextually appropriate meaning. For example, subjects had to make a lexical decision for either *ant* or *spy* upon hearing the word *bug*. There are two meanings of bug, one related to insects and one to surreptitious listening devices. In Swinney's experiment the subject heard the sentence "The man was not surprised when he found several spiders, roaches, and other *bugs*." Just as *bugs* was heard, the subject saw either *ant* or *spy* or a control word, *sew*. If context selects the appropriate meaning of *bug* and if activation spreads only to concepts related to the appropriate meaning, then *ant* should be more quickly accessed than the control word, but *spy* should not. But in the experiment *both* words were accessed faster than the control. Even the word not appropriate for the contextually selected meaning was activated.

The principle that emerges is that all meanings are activated, at least for a brief time. There is also some support for the second implication of the activation theory, namely that the contextually appropriate meaning will "win out" as its activation continues while the inappropriate one decays. When Swinney (1979) delayed the lexical decision by four syllables—i.e., four spoken syllables elapsed between the spoken ambiguous word and the printed lexical decision word—only the word related to the appropriate meaning (*ant* in the example) was processed more quickly. Also, using a different task, Tanenhaus, Lieman, and Seidenberg (1979) found evidence that both meanings of a word were available immediately but after 200 milliseconds only the appropriate meaning was available. It remains possible that if context very heavily biases one meaning over another, even early activation becomes more selective (Simpson, 1981). Overall, however, the conclusion is that semantic encoding proceeds in two phases. In the first phase, lexical access activates all meanings associated with a word. The appropriate meaning receives more activation and is selected for encoding.

This second phase, encoding the appropriate meaning after activation, may be influenced by an ordering principle built into the semantic representation. The model of Hogaboam and Perfetti (1975) assumes that more dominant meanings are selected before less dominant meanings to be tested against the encoded meaning of the context. Their evidence indicates that the dominant (most frequent) meaning of a word is more readily available to the reader than is the subordinate (less frequent) meaning. This ordering assumption can be incorporated into an activation model. We assume that the resting level of activation is higher for the more frequent meaning. Lexical access does affect the activation level of all meanings, but when the alternative meaning is appropriate, the dominant meaning will remain activated for longer than the subordinate meaning.

Propositional Encoding

The encoding of individual words enables the encoding of elementary text units, or propositions. We return to sentence (2) of the story excerpt to illustrate. The propositions of that sentence, *The room was warm and stuffy, so they opened the window,* can be represented as follows:

1. exists (room)
2. warm (room)
3. stuffy (room)
4. exists (window)
5. open (they, window)
6. because (5[2 & 3])

Thus there are six elementary propositions in this sentence. Two, however, are existential propositions that merely assert the existence of some referent. These propositions, in a sense, answer the questions "what room?" and "what window?" that would arise in propositions 2 and 3. No existence proposition is required for *they* because it is anaphoric. It refers to something whose existence has already been established (in the first sentence of the story).

The central element in a proposition is the predicate. Grammatically, it may be a verb, an adjective, or certain conjunctions. It predicates a relationship between two or more concepts in a sentence, as in proposition 5, or it predicates some property or state of a concept, as in propositions 2 and 3. (The existential predicate "to be" can be said to predicate the property of existence.) Proposition 6 is an *embedding* proposition, in that it combines two propositions by embedding them into a third one. It predicates a causal relationship between proposition 5 and propositions 2 and 3.

In short, *propositions* are abstract, elementary meaning units that comprise the meaning of a sentence. A text may be defined as a set of coherently related sentences and may therefore be represented as a list of interrelated propositions. The major theory of text representation based on this assumption is described in Kintsch (1974) and Kintsch and van Dijk (1978).

Working memory

On encountering the words in a sentence and encoding them, the reader assembles them into propositions. Or rather, propositions represent the meaning information that the reader assembles from a sentence. This assembly occurs within a limited-capacity processing mechanism, the same mechanism that can be applied to remembering a string of digits or multiplying pairs of two-digit numbers. Since this working memory (Baddeley & Hitch, 1974; Newell & Simon, 1972) is a limited-capacity mechanism, it follows that the amount of information it can handle is limited. We can assume, with Kintsch and van Dijk (1978), that the reader can hold only a few propositions in working memory at one time. As new propositions are assembled, previously assembled ones are vulnerable to memory loss. The trick, in general, is to quickly integrate the assembled propositions into a representation that can survive in long-term memory. If a sentence is very long, and especially if its end part has a proposition that needs to be integrated with its front part, extra processing effort must be applied.[2]

Thus, reading as a local process is a matter of assembling and integrating propositions for longer term memory. One fact that gives some support to this account is that the time it takes to silently read a short text is a function of the number of propositions, even when number of words is controlled (Kintsch & Keenan, 1973). There are many other important aspects of a model that describes reading in terms of propositions, including exactly how memory is limited, how a representation of the text is built up in memory, etc. These are discussed in Kintsch and van Dijk (1978). Although many processing details are not well under-stood, a central point is that the reader processes some kind of text meaning unit that seems reasonably well approximated by the proposi-tion. In the course of this processing, he constructs, in the ideal case, a coherent set of propositions that constitute the text's meaning.

Integrative processes
Processes that integrate text material are a continuous part of reading. As a local process, *integration* refers to combining successively occurring propositions with each other. For example, in the visit-to-the-doctor story, the six propositions of sentence (2), are integrated with each other and with the representation of sentence (1). The latter we assume is at least temporarily in short-term memory. Thus *local* integrative processes are those that occur by linking recently formed representations with new propositions as they are encoded.

A major mechanism for local integration is the recurrence of an ele-ment in different propositions. Kintsch and van Dijk (1978) refer to this recurrence as *argument overlap* (nouns are "arguments"). For example, the two propositions (2 and 3) in sentence (2) both contain *room*. Ac-cordingly they are quickly integrated in working memory: roughly speaking, the room that is warm is also the room that is stuffy. A some-what more interesting example can be seen in proposition 5 of sentence (2), *they opened the window*. The integration here has to occur by link-ing the pronoun *they* to something earlier. The reader has to link *they* to *Joe* and *his infant daughter*, which are represented in a proposition of sentence (1).

Such a linkage would be carried out first by recognizing that *they* is anaphoric, i.e., having some antecedent element as its reference. Then a search of memory is made for the reference, and *Joe* and *his infant daughter* are quickly found. All this happens quite quickly and automat-ically unless the antecedent is no longer in short-term memory. Thus lo-cal integration is a process that depends on linguistic signals that trigger attachments in memory.

In addition to pronouns, another type of linguistic trigger is the defi-nite article. It tells the reader either to expect a match in memory or to build one. For example, in sentence (2) the reader immediately encoun-ters "the window." But what window? His text memory contains no window, but the definite article *the* is a signal to infer the existence of a window. So proposition 4 establishes the existence of a window. In or-der to do such inferring, the reader must have knowledge about rooms, including the knowledge that they often have windows. The previous mention of *room* has activated the knowledge needed so that a new proposition may refer to windows, floors, furniture, and any of the things that rooms ordinarily have. (This is schema activation, discussed in the next section.)

Of course, a very useful linguistic trigger is the repetition of a word itself. The next mention of "Joe" or "the doctor" or "the window" in the story will immediately lead to a link with a previous proposition containing one of those words. However, the word itself is not the key. Haviland and Clark (1974) demonstrated this with sentences of the following kind: (1) *Marvin liked beer.* (2) *The beer was warm.* The subject who reads these sentences encounters a repetition of the word *beer.* However, the first appearance of the word has not established the existence of a particular reference. (There would be no existential proposition representing the existence of beer.) Thus the linking attempt for the second occurrence of *beer* would fail. Haviland and Clark (1974) found that subjects took longer to read sentence (2) than when it followed a sentence that established reference by mentioning some specific beer.

An important process that occurs as part of local integration is *reinstatement* (Lesgold, Roth, & Curtis, 1979). It is fairly typical for some proposition to be displaced from working memory by subsequent text processing. As we noted, one thing that can happen is an active search of memory for the needed information. Sometimes the information needed is fairly quickly reactivated (or "reinstated"), because it is kept in the foreground by the text that follows. For example, Lesgold et al. compared the time to read a given sentence under several conditions. In one, a necessary antecedent was kept in the foreground. In the other, it slipped to the background. For example, this text mentions *forest* in the first sentence. Then it keeps the forest concept in the foreground, even though it is not referred to again:

> A thick cloud of smoke hung over the forest. The smoke was thick and black and began to fill the clear sky. Up ahead Carol could see a ranger directing traffic to slow down. The forest was on fire.

In the comparison condition the intervening sentences (those not italicized) did not refer to *smoke* nor to *forest ranger,* but to activities inside the car. (What car? The reader's inference is invited.) The difference between keeping the forest in the foreground and letting it slip into the background is that the final sentence (italicized) is read more quickly in the foregrounding condition.

Thus, as Lesgold et al. describe it, there are three integrative processes that can occur when a sentence is read: (1) An immediate match can be found in active memory (or short-term memory), based on recently encoded propositions. (2) When the activated memory does not contain the match, the information is easily *reinstated* when intervening text has kept it foregrounded. (3) When there is no immediate match in active memory and when the required information has not been found, a search of memory is required. Comprehension in such a case will depend on the reader finding the relevant information and perhaps making a "bridging inference" (Haviland & Clark, 1974).

Thus integrative processes depend on linguistic triggers and the accessibility of linking propositions in memory. To the extent that what is accessed is more than the most recently encoded text, we are discussing not merely local processing but the construction by the reader of a model of the text.

THE READER'S TEXT MODEL

There is, of course, more to reading a text than encoding words and propositions. The reader encodes these propositions in the context of knowledge about concepts, knowledge about inferences (inference rules), knowledge about the forms of texts, and general knowledge about the everyday world. By *text modeling* we mean the processes by which the reader combines such knowledge with local processes to form a representation of the text meaning. It is this representation that the reader consults at some later time to recall or to answer questions about what has been read.

Knowledge and Inference

Consider again the story excerpt.

> (1) *Joe and his infant daughter were waiting for the doctor to get back from lunch.*
> (2) *The room was warm and stuffy, so they opened the window.*
> (3) *It didn't help much.*
> (4) *Why wasn't there an air conditioner?*
> (5) *Strange, for this part of Manhattan.*
> (6) *Suddenly the door swung open.*

After reading this excerpt, a reader might be asked to recall it exactly. It is possible that his recall attempt would include the following in place of sentence (2): "Joe opened the window in the office because it was hot and stuffy." The chances are one has to check back with the original just to be sure of what the differences are. First, *hot* is recalled instead of *warm*. This is a local text change based on encoded word meaning. Such a change can be accounted for by reference to semantic memory connections and lack of precision in encoding. However, recalling *office* instead of *room* is a different matter, although it looks quite the same. *Office* and *room* are not so closely related in semantic memory. Instead, the error has occurred because of what the reader knows about the everyday world, i.e., that people visit doctors in places called "offices." (If the text had been about Joe's waiting in a room for a late-arriving airplane, the *office* error would not have occurred.) Finally, the same sort of process causes the reader to recall *Joe* instead of *they*. The reader knows that *infants* are not plausible agents for opening windows and infers that the real agent was the adult. Thus some text-recalls reflect processes of word encoding while others reflect a combination of what the reader knows and what is readily inferred.

Some of the inferences a reader makes are semantically or logically implied by the text propositions (entailment) while others are not. The office inference examples above are in the latter category. Their plausibility is strictly psychological, based on the reader's world knowledge. This means, of course, that they are not guaranteed to be correct inferences. In fact, Joe could well have been in a "waiting room" rather than an office. The *Joe* for *they* case is trickier. It is in fact a logical entailment that if Joe and his daughter *(they)* opened the window, then Joe opened the window. So it is a necessarily true inference. But such recalls, if the reader really remembered *they*, would violate pragmatic conventions, observed by the writer, on saying not only what we know to be true but what we know will not be misunderstood.[3]

For a better example of a logically or semantically forced inference, suppose that a reader recalls something about the window being closed when Joe and his daughter entered the room. It is semantically forced because sentence (2) tells of opening the window. Something can be said to have been opened only if its initial state was closed.

Despite their differences, it is probable that all these inferential alterations are governed by a single comprehension principle: In processing a text, the explicit propositions are combined with the reader's knowledge to produce a text model. This knowledge includes what the reader knows about word meanings, e.g., the relationship between *open* and *closed* and *hot* and *warm* as well as what he knows about doctor's offices and other matters of the world. The organized knowledge about such things is a *schema* (Anderson, Spiro, & Anderson, 1978; Bartlett, 1932; Rumelhart & Ortony, 1977).

Schemata

A *schema* is a conceptual abstraction containing slots (or variables) to be instantiated in various ways; it can apply to simple word concepts as well as doctor's offices and restaurants (a favored example of schema theorists). For example, a schema for window opening would include slots for the manner of opening—by lifting, by pushing, by turning a handle, etc. The core of the schema, its invariant part, has to do with causing a window which is closed to become open. More complex schemata have the same properties.

A visit to a doctor includes variable slots relating to appointments (essential or advisable?), receptionists (is the receptionist also a nurse?), and waiting rooms (*Reader's Digest* or *U.S. News and World Report?*).

In the previous discussion the examples of hypothetical "errors" in recall were partly to demonstrate how normal comprehension works. As we realize that the reader's knowledge about many things is important in comprehension, our theory of reading must take account of such knowledge. We must assume that there is available to the processing mechanism the full range of the reader's knowledge—knowledge of semantic relations, rules for inference, and everyday general knowledge. It is possible to suggest that much relevant knowledge is in the form of schemata, organized knowledge structures. Even some inference rules can be defined, at least partly, as the application of schema knowledge to texts which, as all texts must, leave out some of the information contained in a schema.

An experiment by Anderson, Spiro, and Anderson (1978) demonstrates how schemata organize information for a reader. Subjects read either a passage about having dinner at a fancy restaurant or one about a trip to a supermarket. Eighteen key food items appeared in each of the two passages. The question was whether these 18 items would be recalled equally well by readers of the two passages. The schema hypothesis predicts otherwise, on the assumption that a restaurant schema—or "script" as Schank and Abelson (1977) refer to it—organizes food items more powerfully. For example, the restaurant schema includes an ordering principle for classes of foods (appetizers, soups, salads, entrees, desserts). Restaurant meals also contain fewer food items than supermarkets. In a supermarket, oregano is organized at the same level as ice cream. In a restaurant meal it is not. For at least these reasons, the restaurant schema should enable more recall of food items than the supermarket schema, and this was in fact demonstrated by the experiment.

The contribution of a schema in such a situation is at least partly as a plan for retrieval. Whether a schema is used during comprehension as well as during recall requires a different demonstration. Two examples: Bransford and Johnson (1973) constructed extremely vague passages that were nearly unrecallable by subjects. Their vagueness meant that constructing a text model was very difficult. However, when subjects were given a picture clue or a title prior to reading the passage, their recall was much better. Subjects also rated the passages as more comprehensible.

A second example comes from Dooling and Lachman (1971), who constructed a metaphorical passage in which the hero asserted that "an egg, not a table, correctly typifies this unexplored planet." When subjects were given the title "Christopher Columbus" *before* reading the passage they recalled it quite well. When the title was given *afterward*, it was not much help. Thus, such studies demonstrate that schemata are more than plans for retrieval. They are also structures which serve the comprehension of a text. (See Figure 3-2 for an example of how schemata are important in comprehension. It is a genuine example, not one for experimental purposes.)

Exactly how schemata facilitate comprehension is not quite so clear. As Anderson et al. (1978) put it, a schema serves as a scaffolding on which to construct the meaning of the text. However, there are several things that must happen to make this work. First, the appropriate schema must be activated (or identified). Then the text interpretation is guided by knowledge activated by the schema. For example, the schema directs the reader to understand the passage of Dooling and Lachman as about Christopher Columbus and not about an egg. In addition, as we have seen from the story about the doctor's office, there are inferences to be made and the appropriate schema directs these as well. (These functions of schemata are discussed by Rumelhart & Ortony, 1977; Schank & Abelson, 1977; and others.)

However, there may be equally powerful schematic functions regarding the reader's expectations. Refer back to the doctor's office story. There are only six sentences, but already expectations have been aroused by the doctor's office schema. The reader expects perhaps an examination (of the daughter?) by the doctor. The reader at least expects the doctor to show up and do something with Joe and his daughter. If something in the story does not fit the schema, it may receive extra attention. In a sense, the activation of a powerful schema and a text that conforms closely to the schema at all points may be processed mainly in a top-down manner. The reader can simply confirm what he expects. (This may make for rather uninspiring reading in some cases.)

What happens to information that does not fit the schema? It is not simply filtered out. True, people often recall the gist of a passage, leaving out detail. But they have very good recognition memories for actions not typical for the schema (Bower, Black, & Turner, 1979; Graesser, Gordon, & Sawyer, 1979). Even recall, under certain conditions, shows good memory for atypical schema information (Graesser, 1981). Thus a model needs to account for the reader's better recall of gist, his poorer recall of detail, and his ability to remember atypical information.

LOS ANGELES — The latest fall-out of the space program is an astonishing data-recording system developed by scientist E.T. Seti.

The brain of the machine, located on one end, is known as the Data Stream Imprint System, or DSIS, a NASA-copyrighted graphite linear feed designed to perform a message-recording function. Protection against the hostile environment of space is provided by a cellulose-fiber, reinforced-resin protection layer.

The opposite end of the machine incorporates an ingenious solution to error-correction problems, an abrasive data character erase module. In the erase mode, the module is briskly rubbed across the characters to be deleted and, by the phenomenon known in physics as "lift-off," the undesired characters may easily be expunged.

Reliability tests, conducted under rigid National Aeronautics and Space Administration test parameters, revealed an extremely low failure rate, with the graphite fracturing only once in every 1,000 performances. (A peripheral product has been devised, to be used in the rare event of such a failure, that incorporates the use of a clever device called the Linear Feed Maintenance System. The LFMS literally "sharpens" the recording implement by removing the cellulose-fiber protection layer, exposing a fresh length of graphite.)

When I called Dr. Seti to ask him about the instrument, he conceded that each time it is "sharpened" it shortens, which limits its effective half-life to the operator's ability to handle short stubs. That problem is elegantly solved, however, by the interchangeable "throwaway" concept, enabling the operator to select, at his option, the manual override mode provided. This allows him to discard the stub and replace it with a brand new data-recording instrument from the box of spares provided.

These devices cost NASA $237.50 each. They will be offered to the public at 49 cents, including an instruction booklet, a fully equipped LFMS and a 90-day warranty.

Another spin-off of the space program, soon to be made available to the public, is a portable unitized earthwork synthesis system, sometimes referred to as the terrestrial transport shuttle. This product is designed to relocate dirt among piles, or even dig a hole, when properly manipulated by the operator.

A remarkable design feature is the control stick, better known to NASA as the air-to-ground interface contour adjustment probe, developed by engineering physiologists for optimal utilization by low-skill-level personnel. The stick permits the operator to assume a command and control function over the leveraged tactile-feedback geomass delivery module located on the Earth end of the system.

An ergonomic task analysis, conducted under simulated field conditions and timed by stopwatch, was carried out by an operations research team. As a result, a simple step-by-step program was devised, which can be used by any nonscientist in any interactive man-machine situation requiring the relocation of dirt from one pile to another.

Because the proper use of the device requires a "shoving" action, NASA marketing experts first recommended that the gadget be called a shovel, but later rejected the term as insufficiently descriptive.

A recent inventory of NASA's warehouse uncovered an overrun of nearly 3 million terrestrial transport shuttles on hand, for which Uncle Sam shelled out 1,200 disbursement dollars apiece. The shuttles will be offered to the public on a first-come first-served basis, at a surplus-disposal price of $7.37 each, or two for $9.39 — no refunds, no exchanges, limit two to a customer.

There is still another NASA product of which the limitation of space permits only brief description — the individualized reciprocating bicuspid plaque-level limiter, which will be released to the public under the simple Air Force designation of the unitized chopper disequilibrator.

This sophisticated machine was designed to be used by astronauts during zero-gravity teeth-cleaning missions. A preliminary parametric analysis has indicated that it operates most efficiently when used in conjunction with a viscous detergeable purging agent that the astronaut squeezes out of a compressible tube. Tentative strategy calls for this operation to be followed by a flush-out procedure that involves the use of large drafts of an H_2O dilutant that will be specially created for a weightless and benign environment.

Pricing information on this product will be released at a later date. More data on any of these products may be obtained by contacting NASA's Washington office. Dealers are requested to submit sealed bids for quantity purchases.

Figure 3-2. This excerpt from a newspaper column shows that schemata can be necessary for understanding even "real life" texts, as opposed to those made up by researchers. This one was written by Jack Catran for the *Los Angeles Times* and was reprinted in the international edition of the *New York Herald Tribune,* from which it is excerpted here with permission.

One model that does this has been proposed by Graesser (1981). (Similar models have been proposed by Bower et al., 1979; Norman and Bobrow, 1979; Spiro, 1980; and Thorndyke and Hayes-Roth, 1979.) This model assumes that the memory representation includes a "pointer" to an appropriate schema and a set of tags for elements that are atypical for the schema. Tags are also attached to typical information, but the more typical it is, the less likely it is to be tagged. If this seems to imply that readers will fail to remember the typical schema information, that is true, in a sense. They recall such information very well because it is part of the schema that is activated. They sometimes recall it when it is not actually there. Thus memories for texts (as well as other things) will be a blend of more highly encoded information and more highly inferred information. Both types of information may be part of the reader's text model.

An Example of a Reader's Text Model

Figures 3-3 and 3-4 illustrate the kind of text model that a reader might construct for the doctor's visit story. Figure 3-3 shows successive lines of the text going down and processing activities going from left to right. The processing activities are partly sequential but overlapping. Thus, as the text words are read, propositions are encoded, schemata are activated, and a text model is constructed and updated. Both propositions and schemata are used to construct the text model.

Figure 3-4 shows how the text model might be updated over successive sections of the story. The representation is partly hierarchical, in that the most central component *at the moment* is represented at the top level. While other parts of the model might be hierarchical, they are represented as simple network connections. However, the network is structured also by labeled links. The links correspond to cognitive structures required to understand the content of a discourse. In the case of a story, the underlying cognitive structures are sequences of events that are based on goal-directed activities. Thus *characters*, their *plans*, *obstacles* to their plans, their *actions*, and *consequences* of their actions constitute the central event structure of a story. There are a number of proposals for how to describe these structures (Graesser, 1981; Lichtenstein & Brewer, 1980; Omanson, 1982a; Glowalla & Colonius, 1983; Warren, Nicholas, & Trabasso, 1979). They differ in a number of ways, but share the basic assumption that the structure of events is the central underlying description for narratives.[4] This is what Figure 3-4 represents, although its representation of two-level structure is not typical.

The successive text models of Figure 3-4 can be summarized as follows. After encoding the propositions of the first sentence and having schemata appropriately activated, the reader's text model is that there is

Text Sentence Read →	Propositions Encoded →	Schemata Activated →	Text Model Constructed
1. Joe and his infant daughter were waiting for the doctor to get back from lunch.	1. Joe 2. daughter 3. infant daughter 4. doctor 5. wait (6) 6. return doctor 7. lunch doctor	a. father-daughter schema b. doctor-visit schema c. doctor-lunch schema	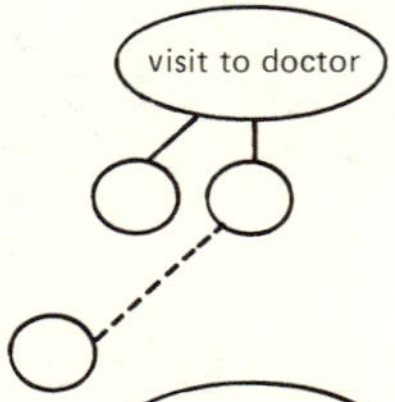
2. The room was warm and stuffy, so they opened the window.	8. room 9. warm room 10. stuffy room 11. window 12. open window 13. cause (12, 8, & 9)	d. window-opening schema	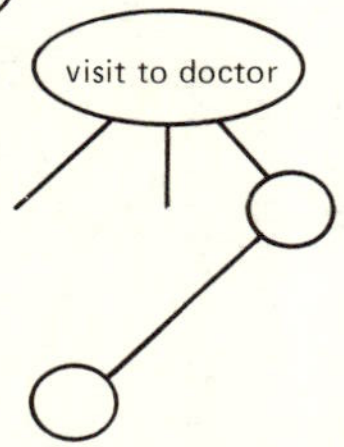
3. It didn't help much.	14. not help (12)		
4. Why wasn't there an air conditioner?	15. no air conditioner 16. why 15?	e. consciousness-of-character convention	
5. Strange for this part of Manhattan.	17. strange (15 & 18) 18. Manhattan	f. Manhattan schema	(see figure 3 - 4)
6. Suddenly the door swung open.	19. open door 20. suddenly (19)	g. room-entering schema	

Figure 3-3. Some parts of the processing of six sentences from the doctor's visit story. Propositions are encoded, schemata activated, and tentative text models constructed. (Propositions are represented in a simplified manner.) This is the order in which these processes occur.

46

Three Successive Text Models

After Sentence 1   After Sentence 2   After Sentence 6

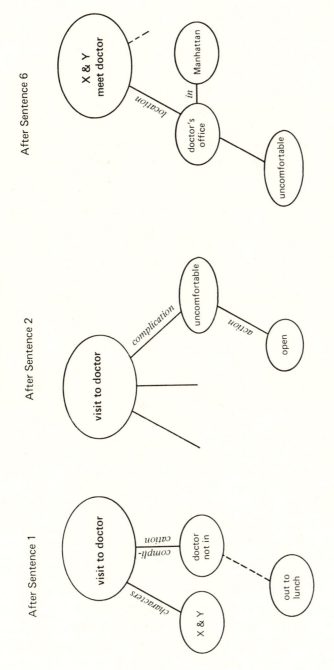

Figure 3-4. Three successive text models in schematic representation. After sentence (2), there is new information, and its potential importance adds it to the Visit head node. The previous links to this node are suppressed for the moment but still available. After sentences (3) and (4), there would be minor changes to reflect the consequence of the action. After sentence (5) (not shown), a location information would be added. After sentence (6) a new model is needed, or soon will be, to accommodate the door opening.

a VISIT TO THE DOCTOR by two people whose goal is blocked for the moment by the doctor's absence. The reader's model is of course very tentative. Encoding the second sentence produces the lower level change. The top-level event is still VISIT TO THE DOCTOR, but the model now represents a second minor complication, the discomfort of the room. Following the propositions of the third sentence, the text model changes only slightly to reflect no consequence of the action taken. Similarly changes following sentences (4) and (5) are minor. They keep the VISIT TO THE DOCTOR as central and add lower level structure.[5] (These models are not shown in Figure 3-4.) The final text model shown is one that the reader might construct following the last sentence. At this point the reader is ready for a switch in event dominance, and VISIT TO THE DOCTOR and MEETING WITH X are now both at the top of the structure. The room-entering is a new event with potential importance. It will quickly recede in importance with more text, either because it was of no consequence (e.g., a mistaken entry) or because its consequence begins to dominate the model (if, say, it is the doctor and he announces a grave diagnosis).

As the story progresses the text model changes, and the *kind* of structures that dominate the model may also change. For example, the *plans* of the characters should begin to dominate their actions and provide the highest level of the text model (Lichtenstein & Brewer, 1980).

Discourse Types

So far, we have focused on knowledge about word meanings and everyday knowledge to explain how a reader constructs a text model. Another relevant sort of knowledge is the kind of text being read. The doctor's visit story is one kind of text type, a narrative. The reader's text model shown in Figures 3-3 and 3-4, that is, one based on event structures, is appropriate for stories. (It is also appropriate for ordinary perception, but that is another matter). In fact, such an event structure is a schema for stories. There are descriptions of story schemata that are complementary to the one described here. These are the story grammars (Rumelhart, 1975; Mandler & Johnson, 1977; Stein & Glenn, 1979; Thorndyke, 1977). A story grammar is a rule-based description of regularities in the structure of stories, predicated on the fact that stories conventionally have settings and episodes. In fact, their structural descriptions are not so different from the kinds of categories appropriate to describe the structure of events. After all, a story is a text about some events that have certain properties.

Exactly what these properties are is an interesting question. Stories do seem to require some aspect of conflict and striving, not a mere haphazard chain of events. However, these aspects can be well captured by event structures that include at least hierarchical properties that subordinate the events to something, especially perhaps a character's plan for action, as Lichtenstein and Brewer (1980) argue. To the extent that story grammars capture features unique to stories as text forms, as opposed to stories as event sequences, there would be an additional source of schematic knowledge for the reader.[6]

Of course, stories are not the only discourse type. There are scientific texts, descriptive newspaper accounts, political essays, letters, etc. The structures of these texts may provide the reader with schemata concerning the distribution of information (Meyer, 1975). For example, newspaper accounts have strong conventions about providing specific information regarding the place, time, and persons involved in a traffic accident or a hold-up. A reader can potentially use such information at least to guide an information search. On the other hand, there is little evidence confirming the use of specific-discourse-type schemata. One exception is scientific journal texts, where there is some evidence that organizational features of those texts are used by readers (Vesonder, 1979).

In addition, there is evidence that readers are sensitive to the linguistic style of a text, which is in fact a matter of discourse type. A legal style, an academic style, and a biblical style are examples that readers can recognize readily. Brewer and Hay (1982) report that subjects' recall of a text reflects the style it was written in, independently of content. For texts written in mixed styles (some parts standard and other parts biblical), recall of standard parts was often in biblical style and vice-versa. Thus it appears that the reader's model of a text will include information about the type of discourse as well as about the content.

SUMMARY

Comprehension in skilled reading includes a number of local processes and text-modeling processes. Local processes are those that operate as the reader gets some meaning out of sentences. They include the encoding of word meanings and the assembly and integration of propositions. Some of these processes will be limited by the functional capacity of working memory. Text-modeling processes combine higher level knowledge and inference processes with the output of local processes to construct a text model.

The encoding of word meanings depends on the highly structured representation of word meanings in memory. The encoding of appropriate meanings is determined by context, but there is a general activation process that briefly affects even meanings not needed by context. As word meanings are encoded, they are assembled into propositions in working memory. There are limits to the amount of information held by this system. Integration of propositions occurs both within sentences and across sentences. Integration is triggered by linguistic devices and can occur as a result of immediate memory matches or reinstatements of deactivated memories or by bridging inferences.

The reader constructs a text model by applying schematic knowledge and inference processes to the results of local processing. These schemata include a wide range of conceptual structures that depend on everyday knowledge, word meanings, and discourse types. The reader's text model is continuously updated by the local processes and the reader's knowledge and expectations.

NOTES

1. Another difference is the basic semantic categories used. These are the labels (above the arrows) to the meaning features in Figure 3-1. There are seven such categories that do not correspond exactly to categories of case grammar nor to any other system. For example, nouns that refer to physically definable objects are *objective*. Their status as locations, human agents, etc., constitutes part of their meaning features. Nouns may also denote *abstraction* (e.g., "understanding," "justice"), *property*, *action*, and *state*. Syntactic modulators are words that are essentially logical operators on concepts, especially the so-called function words.

2. This characterization is fairly general to avoid problems of technical detail. In fact, Kintsch and van Dijk (1978) assume that propositions are immediately chunked into units the size of which is governed by memory limitations and text features. Each unit is a processing cycle. Thus there is already some assembling and integrating, as I have called them. The number of these units that are held simultaneously in memory is a model parameter, usually set at two. As new propositions are encoded they are attached to propositions stored in memory on the basis of argument overlap, i.e., propositions having identical concepts or anaphoric ties. This process occurs at no cost to the limited-capactiy system. There is a cost if a matching proposition is not found in short-term memory. In that case a search of the text base in long-term memory is initiated. Thus, while I speak of assembling and integrating propositions requiring capacity, this does not exactly reflect the assumptions of Kintsch and van Dijk.

3. For example, when asked how much money you have, you would be *truthful* to say "I have $10" when you have $50. If you have $50 you also have $10. But there is a strong constraint on pragmatic grounds (see Grice, 1975), against imprecision. This constraint applies as well to the case here.

4. One difference is whether a model assumes a hierarchical structure above the level of event chains. It is clear that the concept of levels of importance must be captured by some model of the text. There could of course be a level of text model that essentially summarizes event chains, thus

constructing another level. However, the kind of model suggested by Figure 3-4 would directly represent the concept of focal importance.

5. The information in sentence (5) that the scene was in a part of Manhattan where one would ordinarily expect air conditioning is a potentially important clue to future developments. In fact, the reader may immediately sense that a mystery is developing. How *soon* the reader updates the text model will depend on such things. Of course authors of mystery and detective stories depend on such anticipatory devices.

6. There is some evidence that when a nonhierarchical event-based description is compared with a story grammar, each independently can account for subjects' recall (Omanson, 1982b). However, it is possible that this would not be the case for an event theory with a more hierarchical structure. There is some evidence that reading times are affected by story grammar categories (Mandler & Goodman, 1982), but the eye-fixation times of the Just and Carpenter (1980) experiments were unaffected. In either case, the issue is whether the story grammars are powerful schemata only to the extent that they capture event sequences.

CHAPTER FOUR

Speech Processes in Skilled Reading

The role of speech processes in reading might have been treated under lexical access (Chapter 2) or under comprehension (Chapter 3). Or, from the point of view of some reading researchers, it might be omitted altogether. Instead it is here the subject of a separate chapter intended to suggest a theoretical description of speech processes in reading that links speech processes to lexical access *and* comprehension. The proposal is that speech processes occur automatically as part of lexical access and that they support comprehension. But first we will consider some alternative possibilities.

E. B. Huey (1908/1968) was the first psychologist to write a book on reading. He had the idea that private reading is accompanied by a "silent inner voice." Many readers have the same impression that some implicit speech activity accompanies skilled reading. Nevertheless, it has proved difficult even to convincingly demonstrate, let alone to explicate, a role for speech processes in reading. Some have argued that skilled reading is solely a visual and semantic process and that speech plays no serious functional role. These visual theories suggest at most an optional role for speech (Allport, 1977; Barron & Baron, 1977; Davelaar, Coltheart, Besner, & Jonasson, 1978; Frederiksen & Kroll, 1976). The reason for this view is that most researchers have focused on speech recoding as the major speech issue for reading. (See McCusker, Bias, & Hillinger [1981] for a review.)

SPEECH RECODING

It has been common to ask this question: On encountering a printed word, does the reader transform the visual input into a spoken form in order to access its meaning? This is the question of speech recoding. Examples of recoding are actually commonplace. When a child just learning to read encounters the word *cat,* it is not difficult to imagine the word identification process having an intermediate stage: *cat* → /kæt/ → oh! furry Morris! Even adults show recoding, especially in reading a foreign language they have not completely mastered. For example, suppose a reader's German is about good enough to be able to translate *Die Frau ohne Schatten.* In this case, to be "about good enough" means being able to transform the written German into a speech form and then recognize at least part of it. *Frau* → /fraw/ → of course that's *woman!* And *Schatten* /šatən/. If it sounds a bit like *shadow,* that helps quite a bit. Words like *die* and *ohne* are perhaps different—one merely apprehends their syntactic value (if the language is somewhat familiar) just as one apprehends their English syntactic equivalents, *the* and *without.* (For some readers there may also be useful top-down knowledge, namely, that Richard Strauss wrote an opera of this name. This knowledge may make it easier to bypass speech recoding and to translate the phrase immediately into something meaningful.)

The example is important to remind us that there are commonplace occasions in which speech recoding occurs. When reading is new, when a language is incompletely learned, and perhaps when words are unfamiliar, speech recoding does seem to be a process used by the reader. The question for skilled reading, however, is whether such a process is *ordinarily* operative. There are three logical possibilities: Speech recoding does occur, it may occur optionally, or it does not occur. The last possibility is ruled out by common sense. The weight of the evidence appears to be that speech recoding may occur as an optional strategy in skilled reading, but that it will most often not occur. This is the view represented in Figure 4-1.

According to Figure 4-1, a visually presented word is encoded as a string of letters, which are used directly to access the word's meaning in memory. Of course both direct access and speech recoding are much simplified in the figure. For example, the visual representation sufficient for direct access may be some momentary synthesis of the letters generated by the reader (Massaro, 1975). The important fact is that information-processing models with very few exceptions (notably Gough [1972]) assume that speech recoding is optional and ordinarily plays little role in lexical access.

The experimental evidence supporting such a model has come from different tasks. Lexical decision experiments have provided some of the most important results. One conclusion is that the time it takes to decide that a nonword (pseudoword) is in fact not a word is affected by the phonetic value of the nonword. *Brane* is not a word in English, but it can be pronounced exactly like the real word *brain.* That is, it has the same

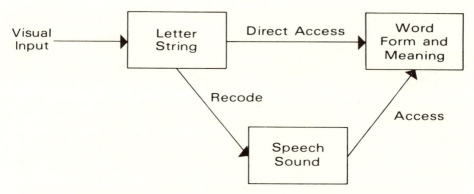

Figure 4-1. The conventional view of direct access versus mediated access. A word's memory location is accessed either from the visually encoded letter string (direct access) or following a recoding of the letter string into sound.

phonetic value. Compared with a pseudoword with equal visual similarity to *brain*, *brane* takes longer to decide. That is, *brane* takes longer than *brone* (Coltheart, 1978). But notice that this is not a matter of lexical access since neither *brane* nor *brone* is a word. (Lexical access, by definition, involves words.) Such a result merely demonstrates that the pseudoword is recoded into speech and that if its sound matches that of a real word, then the subject has to recheck to make sure it is not really a word.

The critical comparison, at least according to Colheart (1978), is between words that are homophones and words that are not. (Homophones are words that have the same phonetic value, such as *groan* and *grown*.) When careful controls for word frequency are made, homophones are not faster in lexical decision than nonhomophones (Coltheart, Davelaar, Jonasson, & Besner, 1977). By one interpretation (Coltheart et al., 1977; also Rubenstein, Lewis, & Rubenstein, 1971), a homophone should be quicker to decide if a word's speech sound is encoded. This is because there would be two chances instead of just one to have a lexical access. So a lexical decision should be faster with a homophone. The fact that it is not is taken to support the idea that recoding does not occur (Coltheart, 1978).

According to one interpretation of the model of Figure 4-1, the reader may attempt lexical access along both routes (Meyer & Ruddy, 1973). This is a "horse race" model; the route that results in lexical access is the "winner." The test for this hypothesis has been the comparison of "regular" words with "exception" words. Regular words are those that have more predictable correspondences between their letters and their phonemes. Exception words are those that are irregular, at least in the sense of having less common correspondences between letters and phonemes. For example *rough* is an exception word relative to *raft* because the phoneme /ʌ/ is usually spelled u (as in *luck*) and the phoneme /f/ is usually spelled *f*. The horse race model predicts that regular words will be accessed more quickly than exception words. This is because for regular words, the speech recoding route sometimes may win as letters are recoded into phonemes. For exception words, the visual route should be the only alternative. Unfortunately, the results of experiments which have compared regular with exception words are mixed. Some show an advantage for regular words (Barron, 1981; Glushko, 1981; Stanovich & Bauer, 1978), but others do not (Bauer & Stanovich, 1980; Coltheart, Besner, Jonasson, & Davelaar, 1979).

Actually, a model proposed by Glushko (1981) helps explain conflicting results. In contrast with the model of Figure 4-1, Glushko suggests that lexical access is visual and that activation of phonological information occurs *after* access. According to Glushko's model, the regularity effect appears just when the "neighbors" of the regular word have the same pronunciation it has. For example, access to *rate* would activate such neighbors as *mate* and *late* that have the same pronunciation. However, if a regular word, such as *save*, is accessed, some of its neighbors, at least *have*, would have a different pronunciation. Thus whether a regular word results in a faster lexical decision than an exception word depends on the extent to which words spelled similarly are pronounced similarly. In fact, this result was obtained by Bauer & Stanovich (1980).

One other line of evidence must be mentioned—the effects of "priming" on lexical decision. The critical case is deciding on, say, *couch* immediately following a decision on *touch*. Although these two words are spelled identically except for the initial letter, they have different pronunciations. Research by Meyer, Schvaneveldt, and Ruddy (1974) showed that lexical decisions in such cases (e.g., the decision *touch*) were slower than decisions on words preceded by other words that shared both their pronunciation and their spelling. However, as with the regularity experiments, the research on this "negative primacy" effect is not consistent (cf. Hillinger, 1980). Furthermore, a clear prediction from this line of reasoning is that there should be a positive primacy effect. For example, *grown* should be accessed more quickly when preceded by *groan* than when preceded by *green*. Evidence again is mixed (Davelaar et al., 1978; Hillinger, 1980). Thus evidence concerning lexical access is inconclusive. Sometimes there seems to be a prior stage of speech recoding and sometimes not.

Some intriguing research on this issue has been reported by Lukatela, Popadic, Ognjenovic, and Turvey (1980), who took advantage of an unusual bi-alphabetic situation: Serbo-Croatian (spoken in Yugoslavia) is written in two different alphabets, Roman and Cyrillic. Each alphabet has some letters which are unique to it. However there are some letters which are the same shape in the two alphabets but have different phonetic values. For example, *p* occurs in both alphabets. It corresponds to /p/ in Roman and /r/ in Cyrillic. Lukatela et al. had subjects make lexical decisions under various conditions. The most important case involved a letter string that had two different pronunciations in the two alphabets but was a word in only *one* of the alphabets. On such a string, subjects' decision times were slowed down compared with a case where the letter string received the same pronunciation in each alphabet. Lukatela et al. attribute this result to a level of phonological processing. The letter string produces one sound in one alphabet and a different sound in the other. The resulting conflict slows decision time. This result seems to implicate some phonological process, but it does not mean that individual letters were recoded prior to access. It suggests an automatic phonetic activation process that cannot be suppressed.

In summary, there is evidence that some speech processes may accompany lexical access. But the consensus is that these processes are not necessary as a stage of processing that precedes lexical access.

COMPREHENSION AND MEMORY

Even if lexical access can occur without a stage of speech recoding, it is possible that subsequent reading processes make use of speech. The most likely candidate is the immediate memory for what has been encoded. In the discussion on comprehension in the last chapter, the immediate or short-term memory for a text was assumed to contain propositions. That is, the focus was on the reader's representation of encoded text meanings. However, there is more to the contents of short-term memory than meaning abstractions. There are also words and perhaps the speech sounds of the words.

In the verbal efficiency theory of reading (see Chapters 6 and 7), there is a special status of immediate memory. In particular, it holds a fairly exact record—within its limitations—of what is read. This means that the words themselves, not merely their meaning abstractions, are available to the reader for a brief period of time. There is an important value to the reader in this verbatim record: It helps secure reference.

Reference Securing

When a reader encounters a written word, lexical access may provide all the meaning information that is needed. Indeed, it is fair to assume that the reader's initial encounter with a word will result in an attempt to immediately encode the word's meaning—not merely its meaning as represented in semantic memory, but also its semantic role in the sentence. For example in sentence (1) and sentence (2) the identical word *theory* has two very different semantic roles.

(1) *The theory explained how reading works.*
(2) *The theory of reading was strongly criticized.*

In sentence (1) theory is an *instrument* of explanation, whereas in sentence (2) it is a *recipient* of criticism. These semantic roles are sometimes referred to as *cases* following the example of case grammars (Fillmore, 1968). These roles are in fact part of the propositions that are encoded.

Thus the *immediacy* assumption (Just & Carpenter, 1980) is that words are fully encoded with respect to basic meaning and with respect to their semantic roles as soon as they are encountered. By this assumption, lexical access results in the complete encoding of a word fully sensitive to its use in the sentence. There will be times when a more tentative encoding occurs or the reader has to reinterpret the encoding in response to information later in the sentence. However, whenever possible the reader immediately tries for a full encoding, according to the immediacy principle.

However, the reference-securing problem is that the initial access to the word may not be sufficient to specify the encoding ultimately needed. For example, consider the following short text:

(3) *John takes Pepe, his dachshund, to football games with him.*
(4) *Believe it or not, they make him buy an extra ticket.*
(5) *They make quite a pair, especially when Pepe jumps on John's shoulder to get a better view of the action.*

In reading *dachshund* in sentence (3), how does the reader encode the word? Semantically, encoding it as [dog] may be sufficient. In fact, later the reader may very well remember only that it was a dog, not that it was a *dachshund*. However, notice that there are interpretations of sentences (4) and (5) that are enhanced if the code is more than [dog]. The anomaly of requiring an extra ticket is that a dachshund is not such a big dog. It would not seem so strange to demand that a husky or St. Bernard pay for a seat. Similarly sentence (5) will be understood differently depending on whether the reader's code is [large dog] or [small dog].

Of course, the semantic encoding can be enriched exactly along these lines, i.e., the code can include semantic category information and essential attribute information. The problem, of course, is how is the reader to anticipate what attributes are essential to encode? In the dachshund example, subsequent text may not need the size of the dog but rather its characteristic coloring, ear size, or personality characteristics.

The *reference-securing principle* is that these potential problems of encoding are solved by a reference-secured code that is part of the reader's activated memory. Such a code allows access to the information stored with a word in memory even after the word has been accessed. This code is the *name* of the concept. Since needed semantic information for a word will be connected in memory by links to the word's name, the name itself will serve nicely as a reference-secured code, regardless of the exact nature of the needed semantic information.

The value of a name-based reference code is even more clearly seen for words that really do need context for their interpretation—for example, most verbs, the more abstract nouns and adjectives, and words whose function is strictly syntactic. Consider again sentence (3). When the verb *takes* occurs, there has not been enough context to secure its reference. *Take*, like most verbs, is very context dependent. Compare, for example, *take a bath* with *take a piece of cake*. In sentence (3) *take* has roughly the meaning of *[x causes y to accompany him to z]*, but it depends critically on encountering the word *with*. This does not happen until six words later. If the word *take* (it's "name") is available in memory over this six-word span, there is no problem and reference is secured.

One interesting feature of the reference-securing hypothesis is that the name code may be incomplete and still be helpful. This is because it is part of a code that has partly redundant components. In the case of the dachshund, the code would be something like (dog, small, /daks . . ./). This would be sufficient to secure the dachshund reference and discriminate it from both poodle and dalmatian. In some cases, merely the initial phoneme of a word is enough to secure the reference. In other cases, more phonetic information is required (see Perfetti & McCutchen, 1982).

Finally, this reference-securing process may apply more typically to words with referential (semantic) content than to words that have only syntactic functions, such as auxiliary verbs and articles, and prepositions (so called "function" words). Such words serve mainly to coordinate words that need to be reference-secured. In fact, the evidence from eye movement research is that syntactic words are fixated only about half as often as content words (Rayner, 1977; Just & Carpenter, 1980). Furthermore there are differences between syntactic words and content words in lexical decisions. Decisions for content words are a function of their printed frequency, whereas decisions for syntactic words are not (Bradley, 1978). It is possible that syntactic function words are not processed by the same mechanism that processes content words. Their reference may be secured only indirectly as part of a referring phrase. However, the main claim of the reference-securing hypothesis is that reference securing is achieved by the activation of phonetic information that allows the securing of a later reference.

A MODEL OF AUTOMATIC SPEECH ACTIVATION

If name codes are useful for reference securing and if phonetic information is part of a name code, then reading is more efficient if phonetic information is part of lexical access. Of course, if a text remains available to the reader, regressive eye fixations can reaccess the word as necessary. But such regressive eye movements do not seem to be very frequent. An efficient system would automatically activate phonetic information during access and use regressive reaccessing only when necessary.

The key process is activation as part of lexical access. Figure 4-2 shows the sketch of the model described in Perfetti and McCutchen (1982) adopted from Rumelhart and McClelland (1981). The question of speech recoding becomes irrelevant in this model. Phonetic activation is not a first step to lexical access; rather, it is part of the access, sometimes reaching a high level prior to the completion of access and sometimes not. The former would look like "recoding" and the latter would not.

The model in Figure 4-2 represents a display of information available over a brief period. It is, in part, an extension of the interactive model described in Chapter 2. There, letters, features, and words were represented at different levels. Figure 4-2 ignores the feature level and shows instead a phoneme level. The horizontal time line represents a brief period, less than a second.

A single word is activated by the lexical access processes described in Chapter 2. Visual information activates elements in memory. The first enclosed box indicates mutual activation of words, letters, and phonemes. As this is a fully interactive model (see Chapter 2), activation flows from the letter level to the word level and vice versa. Equally important, activation flows from the letter level to the phoneme level (and vice versa) *and* from the word level to the phoneme level. This means that as a word is identified, its constituent phonemes are activated. If the identification process is slow, the phoneme level gets a lot of activation from the letter level before a word decision is made. In this case, word identification will be affected by phoneme activation. If the word is identified fairly rapidly, then a phoneme activation will lag behind by a few milliseconds. The eye can move forward to its next fixation during this period.

What is critical is the assumption that some phonetic information linked to the word has been activated by the time the next word is accessed.

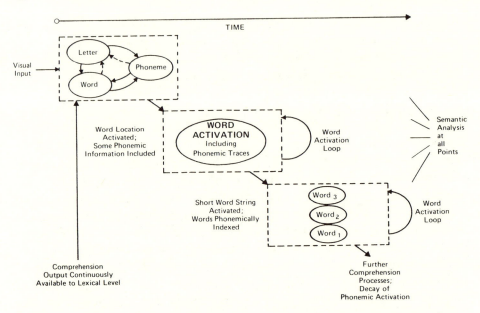

Figure 4-2. Schematic view of an automatic speech-activation model. Visual perception of letters activates both words and phonemes consistent with the current status of letter identification. Following Rumelhart and McClelland (1981), there is feedback (e.g., word-to-letter and word-to-phoneme) as well as feedforward in this process. Automatic phonemic activation occurs during this process and as a result of a word identification decision being made (center box). Thus, a word can be identified before phonemic activation has achieved a high level, but not without some activation occurring. Continuous processing of the word (center box) and reprocessing it as part of a word string (bottom right box) maintain phonemic activation. The processes are continuous (the boxes are not stages) and semantic analysis does not await speech information.

This is represented by the middle frame of Figure 4-2. The word activation loop represents the potential for continuous reprocessing of a word based on its speech sounds. The last frame shows what is activated following two more visual lexical accesses. All words are activated simultaneously and are secured by phonetic information. There are decay functions for this information, and the activation loop helps retard this decay.

Comprehension, in this model, occurs at all points during the process. A word's meaning can be immediately encoded as far as possible. However, there is a memory for the word itself in the form of a phonetic name that can be reaccessed. This name is also part of the propositional encoding in memory. Thus, automatic phonetic activation serves the reader's comprehension. It makes a word available in memory for at least a short time.

Comprehension and Subvocalization

There is evidence for the assumption that comprehension and memory processes are supported by speech sounds. In fact, a general property of short-term memory is that it relies heavily on speech sounds (e.g., Atkinson & Shiffrin, 1968). It may be quite possible for short-term memory to sustain a visual code rather than a speech code (Kroll, Parks, Parkinson, Bieber & Johnson, 1970), and for the congenitally deaf, one must assume a nonspeech coding system. However, except in such special circumstances, short-term memory for both visual and auditory information includes a speech-based code. Perhaps the original experiment of Conrad (1964) demonstrates this fact most simply. When subjects were presented printed letters and then asked to write them down, their errors were based on phonetic similarity more than visual similarity; for example, *F* was confused more with *S* than with *P*, and *V* was confused more with *B* than with *N*.

More relevant for ordinary reading, however, are studies in which the subject does some reading while engaged in tasks that reduce the chances for subvocalization. The logic of such studies is that speech processes, in the form of subvocalization, are helpful to memory, i.e., the reader pronounces the words silently during reading. But subvocalization is not quite what is meant by the assumption that speech processes support memory and comprehension in general. Subvocalization, in fact, may not be typical of skilled reading unless the text is of some difficulty (Hardyck & Petrinovich, 1970).

The research of Hardyck and Petrinovich (1970) and Locke and Fehr (1970), among others, has used measures of electromyographic (EMG) activity (see also McGuigan, 1970; Sokolov, 1972). Electrodes are attached to muscles involved in speech production, and measurements are taken of the amount of speech muscular activity during silent reading. Hardyck and Petrinovich found that the amount of EMG activity from the larynx could be controlled by readers who were given feedback by a bell that rang when their EMG activity increased. When readers kept their laryngeal EMG at a low level, the comprehension suffered for difficult passages but not for easy ones. The speech mechanism appears to be very specific to the sounds that would be produced by subvocalization. Thus Locke and Fehr found labial EMG activity to be increased during processing of words in a memory task with labial consonants, such as *p*, *b*, and *m*.

Such research demonstrates that speech muscle activation may accompany silent reading on some occasions. However, it does not suggest that such activity is characteristic of normal reading, only that it is characteristic of difficult reading or for memory rehearsal. Furthermore, it cannot directly address whether subvocalization activity is functional in the reader's attempt to remember and comprehend or whether it is merely a by-product of this attempt. Research that has addressed this issue has required subjects to produce overt speech while reading. This concurrent vocalization should interfere with subvocalization and, if subvocalization is functional for memory and comprehension, some adverse effects should be seen. The results of this research are quite mixed. Some experiments find concurrent vocalization to reduce comprehension (Kleiman, 1975; Levy, 1975; Slowiaczek & Clifton, 1980) while others do not (Levy, 1978). Still other studies find comprehension accuracy, but not comprehension times, to be affected (Baddeley & Lewis, 1981). The conflicting results are partly a matter of how demanding a given comprehension measure is (Slowiaczek & Clifton, 1980). They are also dependent on the extent to which the concurrent vocalization task makes demands on central processing (Baddeley & Lewis, 1981; Waters, 1981). For example Waters had an assessment of how much a concurrent vocalization task interfered with monitoring a tone (i.e., with a nonreading task). It interfered with reading no more than with the nonreading task. Overall, the important conclusion seems to be this: Silent reading is affected by a reader's concurrent vocalization. It affects both comprehension and memory. But it does so largely because it uses the same limited-capacity working memory that is needed to assemble and integrate propositions. For a more thorough discussion of this see McCutchen and Perfetti (1982) and Perfetti and McCutchen (1982).

This last point, that concurrent vocalization interferes with reading insofar as it uses processing resources, implies one of two things. Either speech processes themselves are not functional in reading or subvocalization is not the relevant speech process in reading. The second conclusion may be the correct one. We have observed previously that subvocalization is not a characteristic part of most reading, although it may accompany difficult reading. Baddeley and Lewis (1981) suggest that subvocalization serves an *articulatory loop,* a rehearsal mechanism of working memory (Baddeley & Hitch, 1974). When concurrent vocalization has an effect on reading it is because it has interrupted the use of this rehearsal loop by the reader. However, according to Baddeley and Lewis (1981), this will not be a problem for ordinary comprehension but only for a process which has to use the rehearsal loop. These include specific memory tasks and other experimental tasks that demand some reprocessing of words.

What then remains of the speech processes that are automatically activated by lexical access? These cannot be the processes of a rehearsal loop because the latter is an optional part of processing. It may be "turned off" for most reading. If we look again at Figure 4-2, we can see the general solution to this problem. The middle part of the Figure 4-2 represents automatic word activation. It happens routinely and makes available to the reader a phonetic code. The lower right part of the figure represents automatic activation of additional words. (Phenomenologically, this activation may produce something like what Baddeley and Lewis (1981) called an "acoustic image.") Additionally, the word activation loop is available for activated words, as shown in Figure 4-2. These loops are optional. Concurrent vocalization can interfere with these loops but apparently not with the initial speech activation. Indeed, if this initial activation is automatic it should be resistant to interference (Schneider & Shiffrin, 1977; Shiffrin & Schneider, 1977).

The relationship between automatic speech activation and other implicit speech processes may be considered as an activation continuum. Under ordinary circumstances, speech mechanisms in reading produce only very pale copies of actual speech. In fact they are not always good copies, because they may include only some of the phonemes, especially the initial phonemes of a word. Under more demanding circumstances, the speech mechanism in reading produces better copies, more similar to actual speech, although of course still unspoken. This general state of affairs can be represented by an activation continuum, as shown in Figure 4-3.

The speech activation continuum represents the degree of activation of a speech mechanism. Normal silent reading activates the mechanism above its resting level. However the representation available at this level is abstract (phoneme based) and impoverished. Increasing activation brings speech motor commands to near threshold level. This is the level at which EMG recordings reveal evidence of speech muscle activation. At a higher level, some of the motor commands get executed and subvocal speech is produced. Actual vocalization may occur at the highest level as voicing commands are produced. By this view, the speech process most used in reading is a lower level of activation that results from lexical access automatically. Vocalization is irrelevant for this activity but not for memory-maintaining activity.

Evidence for speech activation
Since evidence for low levels of speech activation cannot come from concurrent vocalization tasks, such evidence must be sought elsewhere. In fact, evidence for the assumption that speech processes are *automatically* activated is in short supply. However there are some demonstrations that speech processes are activated in silent reading.

One demonstration comes from the visual tongue-twister effect (McCutchen & Perfetti, 1982). This research assumed that the speech codes activated by words have two features consistent with the speech activa-

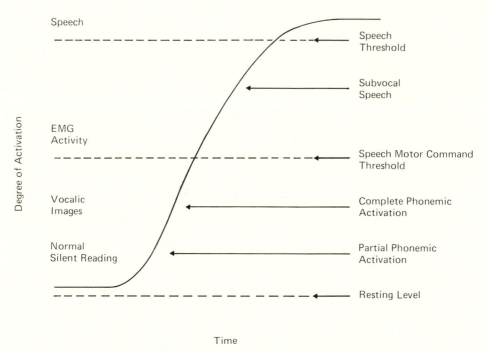

Figure 4-3. Speech activation continuum with thresholds for motor activity. The degree of speech activation increases from a resting level through low levels of phoneme activation through subvocal and true speech. From Perfetti & McCutchen (1982).

tion model discussed in this chapter: (1) They include consonants as well as vowels, and (2) they may be incomplete, in which case they include word-initial phonemes. These assumptions in part reflect the fact that consonants are high in information and that lexical access includes a priority for word beginnings (see Perfetti & McCutchen, 1982). McCutchen and Perfetti further assumed, following the reference-securing assumption, that the distinctiveness of a phonetic code would be threatened by repetition of that code within a sentence. It would take more processing to secure the reference of words that started with the same phoneme. Thus tongue twisters, sentences that repeat initial consonants, e.g., *Peter Piper picked a peck of pickled peppers*, should take longer to read, even in silent reading.

This is what the McCutchen and Perfetti experiments found. The semantic acceptability of sentences such as (1) took longer to decide than sentences such as (2).

(1) *The detective discovered the danger and decided to dig for details.*
(2) *The investigator knew the hazard and chose to search for answers.*

These sentences have nearly identical syntactic structures and comparable semantic interpretations. Thus, the tongue-twister effect is not due to semantics.

The possibility that the effect is visual, i.e., that the reader loses his place during visual processes, is lessened by other manipulations. In one, the tongue-twister effect is found for sentences containing *different* word-initial letters which correspond to phonemes sharing a phonetic feature. For example, both d and t occurred in some sentences (/d/ and /t/ are both alveolars), and b and p both occurred in other sentences (/p/ and /b/ are bilabials). Also in a later experiment with children, the visual factor was reduced by having some words in uppercase and some in lowercase. Thus the tongue-twister effect seems to be due to phonetic activation during silent reading. However, it does not really demonstrate that the activation is automatic.

Another demonstration, perhaps more convincing for the assumption that activation is automatic, comes from Petrick and Potter (1979). They presented words one at a time very rapidly to subjects, at a rate of 12 words per second. Subjects were presented with a probe word immediately following the sentence, with the task of deciding whether it had occurred in the sentence. The key result is for probes that had not actually occurred. When a probe was phonetically similar to a word that had actually occurred, subjects took longer to reject it (and they made more errors). Importantly, this did not happen for probes that were only visually similar to words from the sentence. Thus, the interpretation is that subjects rapidly access automatically activated phonetic codes as well as semantic ones (semantically related probes took even longer to reject). Since the effect of a phonetically related probe occurred within 80 milliseconds of the final word of the sentence, it is plausible that the phonetic activation had taken place when the word in the sentence was accessed. The alternative possibility is that after the sentence is read, phonetic activation occurs as the probe initiates a memory search.

Perhaps the most direct demonstrations that phonetic activation may occur automatically during lexical access come from experiments presenting single words. The Lukatela et al. (1980) research with Cyrillic and Roman alphabets (described earlier) suggests that automatic phonetic activation occurs and cannot be suppressed. In such situations there is no opportunity for a later memory process. Navon and Shimron (1981) took advantage of some interesting characteristics of the Hebrew alphabet to provide such a demonstration. Hebrew marks vowel information by diacritic symbols beneath letters, which are largely consonants. However, except for the first school instruction in reading and for certain traditional works, most Hebrew texts omit the vowel symbols. Consonants carry the needed information, and vowel symbols are essentially redundant.

Thus, it was possible for Navon and Shimron (1981) to compose words in Hebrew script that either had redundant vowel symbols or no vowel symbols. Furthermore, some of the vowel symbols were incorrect, i.e., they could accompany some consonants but not the particular ones they were paired with in the experiment. Within this class of incorrectly paired symbols, some maintained the correct vowel sound despite being wrongly paired, while others did not. The interesting result of Navon and Shimron is that the time for skilled readers to read the word aloud was not affected by its having an incorrect vowel symbol, unless the incorrect symbol did not maintain the right vowel sound. By ruling out a visual explanation of this result, Navon and Shimron (1981) suggest that it reflects an automatic activation of letter sounds. This happens with or without the Hebrew vowel symbol and is interfered with only when the symbol triggers an incorrect sound. Of course, it is possible that *naming* the word, not the lexical access, is what requires the phonetic activation.

The final demonstration involves the backward visual masking of words. Naish (1980) and Perfetti and Bell (unpublished experiments) have demonstrated that phonetic effects occur in backward visual masking. The experimental situation used by Perfetti and Bell is illustrated in Table 4-1. A word is briefly exposed for 33 milliseconds. It is followed immediately by a second word (actually a pseudoword), which is on for 20 milliseconds. This pseudoword is followed by a row of *X*'s. Thus, the pseudoword masks the first word, and the *X*'s (a pattern mask) mask the second word. The subject is to identify the first word. In this context, masking is essentially a process in which one visual display interrupts the processing of an earlier display by erasing its stimulus features. (For theoretical discussions of masking processes see Breitmeyer & Ganz, 1976; and Turvey, 1973).

The critical feature in these experiments is the relation of the pseudoword mask to the word target (e.g., *main*). Because the mask interrupts processing of the target, identification is difficult and subjects make er-

TABLE 4-1 Word-masking
Experiment

Trial event	Duration (in msec)
main	30
MAYN or MARN or CRUB	25
XXXXXXX	until subject responds

rors. But when the mask has many of the same letters as the target (e.g., MARN), identification is much improved compared with the case in which the mask does not share letters with the target (e.g., CRUB). In the Perfetti & Bell experiments, this enhancement of target identification occurs even when the target is in lowercase and the mask is in uppercase. Most important is the fact that identification is also improved if the mask has many of the same phonemes as the target, even if the letters are not the same. For example, when the target word is *quote* and the pseudoword is *KWOAT* there is high phonetic overlap not due entirely to letter overlap. (In fact *quote* and *KWOAT* are pseudohomophones. They could have identical pronunciations.) Identification of the target word is enhanced in these cases, compared with a word sharing only letters. Thus, there is an enhancement of recognition when the mask preserves either the letters or the phonemes of the word. The word's initial access has been accomplished by something like the model shown in Figure 4-2 (see also Figure 2-2). Its phonemes are partly activated when the pseudoword mask interrupts processing. As the masking word is processed, its features are activated. If these features, including phonetic ones, are similar to those already activated by the target word, then these previously activated features continue to receive some activation. Thus, the experiments of Naish (1980) and Perfetti and Bell demonstrate that a word presented too briefly to be easily recognized has its phonemes as well as its graphemes activated. This seems to be automatic phonetic activation during lexical access.

SUMMARY

The question of whether skilled silent reading includes implicit speech processes has been difficult to answer, despite its appearance as one of the earliest theoretical issues of reading. Much research has examined whether a stage of speech recoding occurs prior to lexical access and has seemed to demonstrate that it does not in skilled reading.

On the other hand, there are reasons to assume that speech processes occur as support for memory and comprehension. For example, the reader's need to secure reference is facilitated by having a phonetically indexed name code. This code can be automatically activated as part of lexical access, consistent with an interactive model of lexical access as presented in Chapter 2. The outcome of this activation process is a speech code that is not quite the same as a covert vocalization. Subvocalization may be the result of especially high levels of speech activation in response to memory demands.

There is evidence for speech activation during silent reading from sentence-reading tasks. Evidence from word-reading tasks may more directly demonstrate that activation is automatic.

PART TWO

Individual Differences in Reading Processes

CHAPTER FIVE

Individual Differences in Reading Skill

The preceding chapters have offered a general account of skilled reading. They have assumed an experienced adult reader who handles ordinary texts with comprehension and reasonable speed. In this chapter we address the fact that there are differences in reading skill. We have already seen some differences in the discussion of speed reading (Chapter 2). However, here the focus will be on lower levels of reading skill. We will examine the processing factors that contribute to reading skill differences among adults and especially children by considering the components of skilled reading that have been described previously. That is, we will consider individual differences in lexical access and comprehension.

Reading ability definitions
There are many ways in which people can be said to have different reading abilities. Readers can differ in the speed of their text reading (see Chapter 2), their comprehension, their ability to read aloud, and even their ability to read unfamiliar words. Moreover, what is meant by "reading ability" must depend in part on whether we are referring to children or adults. We expect both children and adults to read with comprehension and with some speed. However, we will be more impressed with a 7-year-old who can read a three-syllable word aloud than with an adult who does the same.

In general, reading ability will be defined here as it has been throughout the previous chapters. Comprehension is the reading ability to be described and partly explained. "Children" refers to elementary school pupils beyond the first grade, especially in grades two through six (ages 7–11). By readers of low ability or low skill,[1] we mean children who score in the normal range of standard IQ measures (above 95) and are one to two grade levels below their grade assignment as measured by a standard test of reading comprehension. Some of these children may fit the definition of dyslexia, while others will not. The issue of how dyslexia differs from our description of low ability will be taken up in Chapter 9.

Adult reading ability is defined in a similar manner, although the tests measuring ability may be called "reading tests" or "verbal aptitude tests" and the "low ability" readers are actually quite good. Many "low ability" adult readers studied in research on adult reading tend to be in college. The relationship between adult reading ability and children's reading ability will also be discussed in Chapter 8.

Comprehension ability in general
The possibilities for ability differences in reading comprehension include the processing factors described in previous chapters: the local processing components of semantic encoding and propositional encoding, and the higher level schema processes especially critical in text modeling. They also include lexical access and the speech processes that accompany lexical access.

There are two complementary approaches to how differences in comprehension abilities arise. To understand these approaches, it is useful to recall the overall picture of comprehension sketched in Chapter 3. The reader must encode words semantically and assemble them into propositions. The encoded propositions must be integrated with locally coherent text memory representations and into a globally coherent model of the text. Among these text-modeling processes, two seem especially important for the overall comprehension product: the application of appropriate schemata and the monitoring of comprehension processes.

Thus, one approach to overall comprehension ability is to focus on these higher level text-modeling processes, especially the reader's use of schemata. The second approach is to focus on the local processes of propositional encoding, including those that contribute to the overall text modeling effort. A reader of high ability can be expected to perform all these processes with skill. A reader of low ability may show less skill in any one or more of these processes. First we consider the possibilities that ability differences in comprehension are dependent on higher level processes and knowledge factors.

SCHEMA (KNOWLEDGE) FACTORS IN COMPREHENSION ABILITY

A person comprehends a text only in relation to what he or she already knows. For evidence of this see Bransford and Johnson (1973) and Dooling and Lachman (1971) among many others (Chapter 2). This is the starting point for the possibility that reading ability differences are a matter of differences in knowledge. The point of departure is that a knowledge- (or schema)-based approach to ability must demonstrate that two individuals will show different comprehension as a result of differences in their knowledge.

One demonstration of this sort comes from an experiment by Anderson, Reynolds, Schallert, and Goetz (1976). They presented the following passage to college readers:

> Rocky slowly got up from the mat, planning his escape. He hesitated a moment and thought. Things were not going well. What bothered him most was being held, especially since the charge against him had been weak. He considered his present situation. The lock that held him was strong but he thought he could break it. He knew, however, that his timing would have to be perfect. Rocky was aware that it was because of his early roughness that he had been penalized so severely—much too severely from his point of view. The situation was becoming frustrating; the pressure had been grinding on him for too long. He was being ridden unmercifully. Rocky was getting angry now. He felt he was ready to make his move. He knew that his success or failure would depend on what he did in the next few seconds.

The Rocky passage can be interpreted in two different ways. Under one interpretation, Rocky is planning an escape from prison. Under the other, he is in the midst of a wrestling match.[2] Anderson et al. found that students from a class in educational psychology tended to understand Rocky as a prisoner while students in weight-lifting class tended to understand Rocky as a wrestler. These differences, found in answers to multiple-choice comprehension questions, demonstrate that individuals with different knowledge (provided in this case by different backgrounds) will understand vaguely worded passages in ways dependent upon that knowledge.

However, the principle is quite general and not dependent on text vagueness. Research by Voss and his colleagues has demonstrated this for individuals who differ in their knowledge of baseball (Chiesi, Spilich, & Voss, 1979; Spilich, Vesonder, Chiesi, & Voss, 1979). In these experiments, subjects were matched in general reading ability and differed only in their knowledge of the rules of baseball. That is, they differed in their knowledge of the game's structure—its action schema—rather than in "baseball trivia." Subjects then listened to a half-inning account of a fictitious baseball game. When they later recalled this half inning, there were important differences between subjects high in baseball knowledge and subjects low in baseball knowledge. High-knowledge subjects recalled more than low-knowledge subjects, and their recalls were qualitatively different. High-knowledge subjects recalled more information about events that were significant for the game itself, i.e., the events that make up the game's essential structure.

This structure, according to the representation of Spilich et al. (1979), consists of information about the setting (e.g., who is playing and who is batting in a given situation) and a goal structure consisting of several levels. The goal structure at the highest level is winning the game. But the interesting levels are those that are essentially subgoals that enable the goal of winning. For example, these subgoals include getting runners on base and advancing them and, when the opposition is at bat, preventing these things. Thus the outcome of a particular pitch, say a single to right field, is interpreted by its effect on one or more of the goal structures. Game actions are important in relation to goal structures.

In the Voss experiments (Spilich et al., 1979), recalls of high-knowledge subjects tended to reflect these goal structures more than did recalls of low-knowledge subjects. These recalls included setting information that was relevant for the goal structure as well as actions that affected the getting and advancing of base runners. Compared with high-knowledge subjects, low-knowledge subjects did not do badly at remembering high-level actions, e.g., how the game ended. And they were good at recalling setting information that was not relevant to the game. The difference was at a sort of intermediate level: remembering actions and setting information that help bring about the important changes in the game situation. To summarize the results of these recall experiments in a way that does only slight injustice to the factual details: A person who knows a lot about baseball is more likely to recall that Iota not only led the league in home runs but delivered a timely single to center sending Sagachito home from third with the tying run. A person who knows a little about baseball is more likely to recall that the game was played in ugly weather and that Iota hit safely.

Given such differences, which we take to reflect differences in how text information gets represented in memory, how do knowledge differences bring this about? The role of schema activation (Chapter 3) seems important here. By having goal structure schemata for baseball, readers (or listeners) will have these structures activated during the processing of the text. This should allow them to build a text model by linking action sequences appropriately. That is, the reader or listener can see immediately the consequences of a game action (e.g., this hit will advance the runner) or a game setting (e.g., with this batter's home run record, this is a potentially important pitch). By contrast, individual's with less knowledge may not immediately see the consequences of a goal-relevant action. They know perfectly well that getting a hit is important, but as Spilich et al. (1979) observe, they may less easily follow the consequence of the hit for the advancement of base runners. The effect of knowledge on comprehension can be described as an effect of schema activation on the encoding of propositions in working memory and on the construction of the text model. Spilich et al., as a result of testing variations of the Kintsch and van Dijk (1978) model and some of their own, conclude that the recall differences between high- and low-knowledge subjects rest on both low-level (propositional) and high-level (text model) encoding.[3]

There is some evidence from the Voss research to support the general processing description given above, at least the assumption that it is the *processing* of the text that is affected by knowledge and not merely the recall of it. Chiesi et al. (1979) studied high and low baseball knowledge subjects in several tasks aimed at processing differences. Among other things, they found that high-knowledge subjects needed less information to decide whether a baseball description was one they had heard. This suggests that high-knowledge subjects used their knowledge to construct a more accessible event structure. They also found that, in recognition memory, high-knowledge subjects were better than low-knowledge subjects at detecting changes made in game descriptions. But their advantage was especially large for detecting changes at a high level of the goal structure. Again, high-knowledge subjects seem to show a sensitivity to the structure of the game and apply it to processing.

Schemata in General Reading Ability

The studies described above, and a number of others not discussed here, demonstrate the importance of knowledge of specific content for comprehension. Thus, when texts are not specific in reference, interpretation is guided by the reader's experience (Anderson et al., 1976). And when the text is specific, the quantity and quality of comprehension is restricted by the reader's experience. These are not experiences that produce reading ability in general; rather they produce the means for understanding certain types of text. Indeed, general reading ability is controlled in such studies.

How does this principle—that comprehension depends on knowledge—become applied to the case of general reading ability? By supposing that individuals who are high in general reading ability have more useful knowledge in more situations than do readers of low ability.

As far as it goes, this conclusion must be correct. The more one knows about everything, the more one will be able to read anything. Reading ability then becomes nondistinct from general intellectual ability. In fact, there are very large correlations between measures of intelligence, especially verbal intelligence, and measures of reading ability. There is, in fact, reason to describe verbal intelligence in much the same terms as reading ability for adults as well as children (see Perfetti, 1983). However, it is not obvious that switching the burden of explanation from reading to verbal intelligence is going to make it easier to understand things. This is especially true when practical implications are brought into the picture: Does the importance of knowledge for reading comprehension imply that teaching should focus on getting more schemata into the heads of more children? Actually, that is a reasonable sort of objective. However, the child may want to do some reading in the meantime.

Actually, there is more to knowledge than the mere having of it. Suppose, for example, that two individuals can each be said to have a schema for a baseball game, i.e., each knows the essential structure of the game in the sense defined in the Voss research. However, suppose that one of the individuals has actively participated in relevant goal structure activity, e.g., managing (even a Little League team) or playing simulated games. These two individuals might have equivalent schemata in the minimal structural sense, but not necessarily in their elaborated structures. For example, the "manager" may have a much richer set of connections to specific subgoal structures. These connections would represent strategies for advancing base runners that depend on such things as the number of outs, the prospects of the next hitter, the speed of the base runner, the bunting ability of the present hitter, as well as the score and the inning. It is easy to imagine that there are consequences of such an enriched network for processing. Particular actions can be encoded not only in relation to the goal structure but in relation to typical event structures and their outcomes (e.g., How typical is a bunt in this situation?). Such processing would be described, for example, in terms of Graesser's (1981) schema pointer-plus-tag model (Chapter 3). However, the important point is that individuals can differ not just in knowledge necessary for understanding, nor even just in the presence of basic goal structure knowledge, but also in the richness or degree of elaboration of their schemata. There is little specific evidence on how this level of schematic richness may affect reading, but there is suggestive evidence for some problem-solving activities.[4]

While schema richness as well as basic schema availability may contribute to comprehension, this possibility has pushed us even further from general reading ability. It suggests the usefulness of even more specialized (or more expert) schemata. So the problem of generality remains. It is possible that we should think of individual differences in comprehension ability to be simply the aggregate of individual differences in specialized knowledge structures. This suggests that one person may be able to comprehend an account of a baseball game but not a magazine article about an Italian parliamentary election (certainly an interesting structure of its own), while another person can do only the reverse, and so on. Thus the low-ability reader is one who has too few useful schemata to be able to comprehend very many texts. This is not a very useful concept of reading ability.

There are at least two ways that schema theory can be applied to general reading ability instead of the reading of specific texts. Suppose a reader has acquired a sufficient number of useful schemata, but for some reason the right schema for a given text is not activated. Spiro (1980) has referred to this problem as one of schema *selection*. For example, if a reader of the newspaper article in Figure 3-2 selects only a spaceflight schema (presumably a very inexpert choice), the text will certainly be difficult to understand. Something in the text itself must trigger some other knowledge, at which point the rest of the text becomes quite clear.

However, it is not easy to see how schema selection can become a serious problem. Texts are not ordinarily so perverse as to invite the wrong schema. If schema selection poses a general problem for some low-ability readers, the difficulty probably occurs at a far more local level. For example, in reading the doctor's visit story (Chapter 3), a failure for the doctor's visit schema to be activated could cause some problems by preventing certain critical inferences from being made. But it remains to be demonstrated that such schema *activation*, independent of having the schema in the first place, is a problem for readers of low ability. Furthermore, any such activation problem may be a result of lexical access difficulty.

A second way in which a schema problem could apply to general reading ability may be called *schema flexibility*. It is really a variation on the schema selection problem, because it refers to a reader who fails to accommodate schematic knowledge to specific text information or fails to apply a different schema when necessary. This seems to be what Spiro (1980) referred to as *schema instantiation* and *refinement*. In an extreme form, this problem would be represented by a reader who recalled the *Odyssey* only as a travel adventure involving a faithful-wife—suffering-husband schema. His schema-driven comprehension would have failed to take account of certain atypical vengeance motifs that get played out at the end of the story.[5]

Other examples of a schema flexibility problem might suggest that a reader can be too "top-down" in his comprehension, failing to incorporate information that deviates from the activated schema. This, of course, would severely limit learning. Consequently, we are again faced with a low-ability reader who does not know very much. In any case, a demonstration that this kind of schematic problem is characteristic of low-ability readers would be necessary. Even then, it would not be clear that a failure to instantiate and modify a schema would be very different from "not understanding sentences" or "failing to connect propositions" or other ways of describing more local text problems.

Nevertheless, when schemata are seen as structures that readers must not only "have" but also use appropriately, it is possible to see the important principle of knowledge application. There is no reason to assume that people routinely apply knowledge appropriately. The application of schemata to a theory of reading ability may turn out not to be a matter of having schemata or even selecting the right one instead of the wrong one, but rather of knowing that there are control processes to apply to texts that help get appropriate schemata activated. This kind of knowledge is usually referred to as metacognition (Flavell, 1976).

Metacognition

Children apparently fail at a wide variety of problem-solving tasks because they do not, or cannot, assess their own knowledge or their progress on the task (Brown, 1978). They fail to recognize an increase in difficulty level, to plan ahead, or to monitor the outcomes of their performance. A similar metacognitive failure may contribute to low ability in reading. At minimum, metacognitive abilities must contribute to the development of reading skill as children acquire strategies for comprehension (Brown, 1980).

There are at least three different sorts of metacognitive ability that can be described in terms of reading awareness: (1) awareness of strategies to apply to text comprehension, e.g., allocation of processing resources; (2) awareness of structural levels in a text, i.e., knowing what is important; (3) awareness of local text inference demands, e.g., noticing when contradictory sentences are encountered.

Strategies for comprehension may include a number of controlling activities during reading. For example, reading for a recall test may encourage a different strategy from reading in order to verify a specific fact. However, the strategy most salient for reading ability is the allocation of processing resources. Do low-ability readers spend too much time on unimportant information at the expense of important information? This means that the question of strategy awareness (1) essentially merges with the question of awareness of text importance (2).

As for awareness of structural levels, it is reasonable to expect that important text material should receive more processing resources than less important material. However, this does not always translate into spending more time on important material. An experiment by Britton, Meyer, Simpson, Holdredge, and Curry (1979) found no effect of a passage's importance on its reading time, but a study by Cirilo and Foss (1980) did. Britton, Meyer, Hodge, and Glynn (1979) point out that text features can make the importance of individual sentences more or less salient and that such features may be present in narratives (used by Cirilo and Foss, 1980) more than in expository texts (used by Britton et al., 1979). However, in their eye movement research, Carpenter and Just (1981; Just & Carpenter,

1980) found that longer fixations were accumulated over sentences that related structurally important information in expository texts, especially definitional and causal information. Thus while the effects of importance may be small relative to other text features, reading is affected by text importance under some conditions.

It is possible that low-ability readers behave differently in this regard. In one study, skilled readers' recall of stories reflected the importance of the text content (as rated by judges) more than did the recall of less skilled readers (Smiley, Oakley, Worthen, Campione, & Brown, 1977). However, this difference was largely one between the number of different importance levels distinguished in their recalls. Even low-ability readers recalled more important than unimportant information. This last result was also obtained when the texts of Smiley et al. (1977) were analyzed by an independent system of importance assignment (Omanson, 1982a).

In a study of high- and low-ability adults, Omanson and Malamut (1980) found that the two groups equally differentiated central content from noncentral content (or important from less important). Beck, Omanson, and McKeown (1983) found the same result with third grade readers: No ability differences in distinguishing the central content of a story from its supportive detail. Thus it appears that while, by some measures, low-ability readers may be sensitive to fewer levels of importance, they do have some appreciation of relative text importance.

The third metacognitive ability is comprehension monitoring (awareness of local text inference demands). It would be surprising if there were not difficulties in monitoring for low-ability readers. Even skilled adults sometimes are oblivious to blatant inconsistencies in texts (Wilkinson, Epstein, Glenberg, & Morse, 1980). There seems to be little direct evidence on this kind of monitoring process in low-ability readers. However, there is evidence that some *simple* monitoring skills can be trained in nonreading tasks and transferred to reading. Brown, Campione, and Barclay (1979) found that retarded children could learn to use a general self-checking monitoring procedure to improve their recall of simple stories. While this does not demonstrate a general monitoring problem for low-ability readers, it does demonstrate the importance of having such a skill even for simple reading. The skill here is not awareness of logical connections but awareness and use of simple memory-checking routines.

LOCAL PROCESSING FACTORS IN COMPREHENSION ABILITY

A second source of individual differences in reading comprehension is in the local text processes. These local processes include those processes that enable propositional encoding and integration. Individual differences in these processes can lead to differences in reading ability either directly or through their effect on higher level processes.

Propositions, Sentences, and Working Memory

The major local processing factor is that the assembling and integrating of propositions occur in a limited-capacity processing system. We again will use *working memory* as a means of referring to the part of this system responsible for comprehension of sentences; i.e., local processing. The possibility for reading ability is that low-ability readers have a reduced working memory capacity.

To make clear how working memory differences could produce differences in reading abilities, refer again to the story of the visit to the doctor's office:

> (1) *Joe and his infant daughter were waiting for the doctor to get back from lunch. (2) The room was warm and stuffy, so they opened the window. (3) It didn't help much. (4) Why wasn't there an air conditioner? (5) Strange for this part of Manhattan. (6) Suddenly the door swung open.*

Recall that the sentence about opening the window (2) contains only 11 words, but 6 propositions. Three ordinary sentences can have a dozen or more propositions. Since working memory is limited in capacity, the ability to keep enough information active is critical. If this cannot be done, the construction of local propositional meaning is at risk.

Do low-ability readers have less memory capacity than high-ability readers? Apparently they do. Take first the case of children readers. Perfetti and Goldman (1976) had children listen to stories for comprehension. However, their listening was interrupted by the occurrence of a probe word from the text they had just heard. They were to produce the word that had followed it in the text. Notice that this simple memory task virtually eliminated any demands of production, a skill we might expect low-ability readers not to have. Nevertheless, low-ability readers were less likely to recall the word following the probe. This was especially true after an entire clause of four or five words had intervened before the memory probe.

Notice that these data are for listening. They suggest that any working memory difference is modality free, not dependent on reading. In fact, exactly the same differences are found in reading. When Goldman, Hogaboam, Bell, and Perfetti (1980) applied this probe memory task to reading instead of listening, high-ability readers recalled more than low-ability readers. Again, the advantage of the high-ability reader increased as the number of intervening words increased. For example, in Goldman et al. (1980; Experiment 1), tests occurred following 2 intervening content words (2–6 words total) and following 5 content words (6–12 words total). Skilled readers in third and fourth grade were more likely to recall the target word than less skilled readers at both intervals. However, skill differences in recall were not large at the shorter interval for fourth-grade subjects. Interestingly, for both skilled and less skilled subjects the occurence of a sentence boundary affected recall. When the target word was from the previous sentence, recall was reduced, especially for the longer probe interval. This was true for readers of high and low skill, the same result that was found for listening (Perfetti & Goldman, 1976).

Overall the experiments of Perfetti and Goldman (1976) and Goldman et al. (1980) support the following picture: During both reading and listening to a text, skilled readers in the third, fourth, and fifth grades have a better verbatim representation of the local text than do less skilled readers. They can remember words better from the sentence they are currently processing and from the previous sentence. Their advantage seems to increase somewhat as more processing intervenes, although for younger readers (third grade) there seemed to be a more constant advantage. Finally, this memory advantage does not depend on different strategies for the use of sentence boundaries. Less able readers, as much as more able readers, have better verbatim records of the currently processed sentence, provided the sentence is not too long.[6]

The implication of working memory differences for comprehension ability differences is quite clear. Low-ability readers will be less likely to achieve an adequate level of comprehension. This can show itself in lower text recall or in poor performance on comprehension questions. In fact, these two measures are highly correlated. For example, Berger and Perfetti (1977) found that the low-ability readers had lower quality recalls, i.e., fewer correct propositions, and answered fewer comprehension questions correctly. In principle, recall or question answering could reflect the operation of plans for retrieval or means of accessing information. However, the force of the memory data is that they instead reflect differences in processing. Low-ability readers have less of the text represented immediately at the level of local processing. They are, in effect, assembling and integrating fewer propositions.

Functional working memory
This processing difference at the local level may be described as due to functional working memory. Such a description does not mean that these are differences in "memory capacity," for the meaning of the latter term depends upon measurement operations that are related to some theory of the properties and structures of short-term memory. There are non-

reading situations in which less able readers show poorer immediate memory performance. In some studies, measures of digit span are related to reading ability (e.g., Guthrie, Goldberg, & Finucci, 1972), although there are other studies in which this relationship is not found (e.g., Valtin, 1973).

The subjects of the Perfetti and Goldman (1976) experiment described previously were given a different sort of memory task. This is the probe digit procedure in which a series of digits is heard followed by a digit probe from the list. The subject has to produce the digit that had followed. This task, which was used to estimate primary memory by Waugh and Norman (1965), did not produce differences between skilled and less skilled readers. Thus, their kind of memory problem is not a simple one of primary memory, understood as a simple passive device that holds a limited amount of information. Rather the memory differences are better understood as differences of an active, functional working-memory system. By this interpretation any verbal memory task that requires an active mechanism will produce differences between high- and low-ability readers. (These different understandings of how memory systems can function in reading ability are discussed more thoroughly in Perfetti & Lesgold, 1977 and Lesgold & Perfetti, 1978.[7])

One implication of the working-memory factor is that active memory processing, when added to the stored contents of working memory, will be a problem for low-ability readers. Experiments by Kail, Chi, Ingram, and Danner (1977) and Kail and Marshall (1978) provide some support for this view. Third- and fourth-grade subjects had to store one, two, or three unrelated sentences in memory and then answer a yes-no question that could be answered by one of the sentences. Low-ability readers were more affected by the *number* of sentences stored than were high-ability readers, at least for the time to answer a *no* question. Thus the rate of search through short-term memory may be slower for low-ability readers.

Working-memory differences have also been demonstrated for adult reading ability. Daneman and Carpenter (1980) instructed subjects to read lists of sentences aloud while trying to recall the final word of each sentence in order. While they were trying to remember the final words of sentences 1 and 2, they were also reading sentence 3. In this situation, much as in normal reading, working memory is asked to store some information while it processes other information. Daneman and Carpenter found that a span measure, the number of sentences before the subjects' recall fell below criterion, correlated with comprehension of short passages. It also correlated with verbal SAT scores.

The Daneman and Carpenter experiment also demonstrates a particular comprehension situation that relies on a large functional working memory. Subjects had to indicate what a pronoun referred to in a text. The pronoun was separated from its referent by various amounts of intervening text. Performance on this task was generally a function of how much text had intervened. However, readers with the highest memory spans, as measured by the final-word memory task, were unaffected by how much text had intervened. Finally, Daneman and Carpenter found that memory span correlated highly with listening comprehension as well as reading comprehension. Thus, for both children and adults, comprehension ability, whether through reading or listening, is related to functional working-memory capacity.

Semantic Encoding

Another local process that may produce individual differences is the encoding of word meanings. There are at least two components to this encoding: (1) the availability of a semantic entry in memory, and (2) the encoding of a word meaning appropriate for the context. The first of these is partly the question of size of vocabulary. The facts are as one would expect: Low-ability readers have smaller vocabularies than high-ability readers.

What does this mean for comprehension? A low-ability reader is less likely to have a semantic entry for any given word in a text. If a text contains many unfamiliar words, certainly the reader's comprehension is at risk. Of course, a word meaning can be inferred from context, and as Sternberg, Powell, and Kaye (1983) have demonstrated, this is a powerful ability, especially of highly verbal individuals. Such readers can use morphological clues (such as the presence of verb inflection) and semantic context cues (such as knowing what kind of thing an unknown word must refer to in a certain context). Thus there are two potentially important vocabulary factors related to comprehension ability: (1) High-ability readers know more word meanings than low-ability readers, and (2) high-ability readers are better at inferring the meanings of unfamiliar words in context.

In fact, standard tests of children's reading comprehension include both of these components. For example, the reading comprehension subtest of the Metropolitan Achievement Test contains paragraphs that the child first reads and then answers written questions about. One of the paragraphs is about a person who collects picture postcards and ends with this sentence:

> *After arranging all his cards carefully, Ted puts them in a large scrapbook and writes beside each one the name of the place it came from.*

One of the questions the child answers is this one: "The word *arranging* in the last sentence means (a) looking at, (b) putting in order, (c) list-

ing, (d) settling." There are items such as this one in many reading comprehension tests. Thus it is fair to say not only that the ability to infer a word's meaning from context contributes to reading ability but that this ability is part of the definition of reading ability. It contributes directly to reading comprehension scores. However, notice that such test items can be answered simply by *knowing* what the word means. In the test example, the reader who knows what *arranging* means does not so much infer its meaning from context as verify it.

The second important semantic encoding process is encoding the meaning of a word appropriate for the context. According to the autonomy principle (Chapter 3), this selection process is preceded by a brief activation of all meanings of the word, whether or not they prove to be appropriate for context. Is there an ability difference in this selection process? For an example we can refer again to the reading comprehension subtest of the Metropolitan Achievement Test, which includes a paragraph containing this sentence:

> *Instead of the large, noisy vehicles that crowd our streets, there were horse-drawn carts and carriages.*

The test questions include this one: "In this story *crowd* could be changed to *(a) many people, (b) push, (c) gang, (d) jam."* It is possible that any one of these choices might be correct for some particular context, but this particular text demands *jam,* partly on syntactic grounds and partly on semantic grounds. Unlike the *arranging* example, the question cannot be answered only by the reader's knowledge of the word's meaning. Context must be used.

Thus, the ability to select the contextually appropriate sense of a word is defined by some tests as one aspect of reading comprehension. However, the processing question is whether less able readers tend not to select the contextually appropriate sense during reading. Evidence for this possibility is sparce, although a study by Merrill, Sperber, and McCauley (1981) seems to support it. In this study, fifth-grade children read sentences in which context emphasized certain attributes connected with a key word. For example, in *The girl fought the cat,* some attributes of a cat (e.g., *claw*) are emphasized over others (e.g., *fur*). Immediately after reading the sentence (one second) a single printed word was displayed in one of several possible colors. The subjects had to name the color of this word. (In such cases naming times are normally slower when the meaning of a word is related to the meaning of the sentence just read [Conrad, 1974].) The interesting result was that, for skilled readers, this inhibition effect was obtained with words naming emphasized attributes (*claw*) but not with words naming nonemphasized attributes (*fur*). However, for low-ability readers, the inhibition effect was found for both emphasized and nonemphasized words. Skilled subjects were apparently locked into the contextually relevant encoding of the sentence within one second after reading it. Low-ability readers, by one interpretation, may have a less well-integrated sentence encoding by that time. For them, more of the word's general semantic features are still activated. One might also interpret this result as showing that high-ability readers are more likely to have an appropriate schema activated during reading of the sentence. However, since the best evidence is that schema activation does *not* inhibit activation of inappropriate meanings in skilled reading, this interpretation is less supportable. The tentative conclusion is that high-ability readers may have a more rapid *dampening* process. That is, for high-ability readers, the activation of inappropriate meanings is more quickly suppressed or dampened following initial activation.

Individual differences in word knowledge

Whatever ability differences there might be in processing word meanings in context, there are differences in the knowledge of word meanings that are not dependent on context. High-ability readers know more word meanings than do low-ability readers. However, in addition to the mere number of semantic entries high-ability readers have, their entries may have more semantic detail. In terms of a semantic memory representation, these two factors are the number of nodes (semantic entries) and the number and type of links. Another way to distinguish these factors is by referring to *breadth* versus *depth* of meaning (Anderson & Freebody, 1979).[8]

To understand how both breadth and depth may play a role in reading ability, we consider a study of college students by Curtis (1981). (See also Curtis & Glaser, 1983). The subjects were classified as high or low in vocabulary knowledge on the basis of a multiple-choice vocabulary test. Based on the results of this test, words were classified according to how well known their meanings were: *known* words (95% of subjects correct), *discriminating* words (50% correct, with high-scoring subjects correct more often than low-scoring subjects) and *unknown* words (an average of 28% correct, with no discrimination between high- and low-scoring subjects). In a later test, subjects were asked to define each of these words and encouraged to provide any vague association they might have for any word whose meaning was unknown to them. The question of interest is whether high-vocabulary subjects had both more breadth and depth than low-vocabulary subjects. For example, depth would be implicated if high subjects had more precise definitions of words for which low subjects could give only vague associations. However, Curtis (1981) found that for discriminating and unknown words, low-vocabulary subjects usually could provide no associations related to the word's meaning. By contrast, high-ability readers did have such vague associations. For example, the definition "*desist* is like cease and desist," given by one high-knowledge subject, indicates at least a vague ability to use the word (based, we might say, on context-specific knowledge, not abstract semantic knowledge). Low-vocabulary subjects did not do even this much, producing instead unrelated associations or none at all.

Such a difference is one of *breadth*. High-ability readers may have meanings, no matter how vague, for more words than low-ability readers. It is possible to see the value of such knowledge for getting meaning from context. With some knowledge, even if it is specific context-dependent knowledge more than semantic knowledge, a contextually appropriate meaning can be constructed.

All of this is not to say that high-ability subjects have their vocabulary advantage only in breadth. In fact, Curtis (1981) found that there were differences in *depth* of knowledge as well. Given that a high-vocabulary and a low-vocabulary subject each knew something about a word, a high-vocabulary subject was more likely to be able to give an acceptable synonym or definition. The implication is that both breadth and depth may be factors in a person's reading ability. Knowing at least a little about more words (breadth) may enable contextual processes to operate, and knowing more about some words (depth) may help secure more precise representations.

Word knowledge in comprehension

If there are individual differences in the breadth and depth of semantic word knowledge, how do they produce differences in general reading ability? The most straightforward possibility is that knowing word meanings enables a reader to encode the propositions needed to construct a text model. Comprehension is impeded by not knowing enough of the words.

An experiment by Beck, Perfetti, and McKeown (1982) illustrates the problem. The researchers constructed texts consisting of some well-known words and some words that were largely unfamiliar to the fourth-grade subjects. When about 9% of the words were unfamiliar, recall of the semantic content units of the text (Omanson, 1982a) was only about 19%. Subjects who had been taught the meanings of these words, but who had had no other form of special instruction, were able to recall about 27%. There was nothing particular about the words' place in the text structure. Had they been from the more important propositions, perhaps the differences would have been larger. Nevertheless, the difference is quite enough to demonstrate that text recall depends on knowing word meanings.

More specifically, knowing word meanings enables the reader to assemble and integrate propositions. When words are not known, the reader's initial text representation is incomplete, and the integration of the text is especially difficult. Omanson, Beck, McKeown, and Perfetti (in press) analyzed the recall data of Beck et al. (1982) according to the text-processing model of Kintsch and van Dijk (1978). The question was what kind of processing assumptions would best describe what happened when a child read a text with unknown words. The processing assumption that seemed to provide the best account of the data was this: Readers held the word as part of a proposition in working memory with an incomplete representation. As new text was processed, such propositions needed for linking previous text with new text were less available. The reader instead linked new text with what was available, often a fragment of what was presented. The reader tried to maintain the important information and did so to the extent possible. The major differences between such readers and those who knew the word meanings centered on the ability to recall text at all levels. All readers clearly tried to construct a coherent text model.

Thus, it is possible to sketch a general picture of how individual differences in knowledge of word meanings can produce differences in reading ability. However, an explanation of general reading ability based on such differences has the same problems as a schema explanation. It has to mean that on the average, low-ability readers are more likely to encounter texts with more unknown words. What about a text in which all the words are known? In fact, reading-ability differences are present for such texts, as will be shown in the next chapter. To that extent, vocabulary differences can be said to make an important contribution to reading ability, but not necessarily a decisive one.

LEXICAL PROCESSES

The fact that the encoding of word meanings is a lexical process suggests a more general question: To what extent can individual differences in reading comprehension ability be attributed to differences in processes of lexical access? The theoretical assumption we make is that when a word is accessed, activation of different components occurs. There are letters, word concepts, and word names, these last including speech information, as we emphasized in Chapter 4. So there are differences that could operate at all these levels. First we need an additional perspective on how lexical access works.

Lexical Access and Rule-Based Knowledge

In the discussion of lexical access in skilled reading, the assumption was that access was achieved by interactive activation processes (Chapter 2). This is a convenient way, following Rumelhart and McClelland (1981), to describe lexical access for skilled reading, but it may need some modification in order to apply to individual differences. What it needs most is a way to represent the knowledge that readers have about orthographic principles.

Orthographic structure is a feature of alphabetic writing systems. Understood broadly enough, it is also a feature of any writing system that constrains the co-occurrence of its basic graphic constituents. The orthographic structure of English is such that the arrangement of the graphemes is constrained. In short, there are rules of orthography, many of which have been described in detail (Venezky, 1970).

Let us be clear about why orthographic rules are potentially significant. Insofar as such rules generate (describe) the set of English word spellings, or at least part of that set, they summarize an important sort of knowledge for the reader. The representation of this knowledge is an issue, but any of several possibilities is sufficient for the present purpose. The rules simply have to represent certain facts about graphic constraints. To take the most familiar example, *q* is practically always followed by *u* in English. For another, consonant strings of more than four are very rare. For another, *ch* is common while *hc* is impossible. Many of the rules are context sensitive: For example, *tr* can begin an English word but not end it. Of course there are letter strings that are more rare than illegal and these exceptions to spelling "rules" are widely noted.

There are different sources for the spelling rules of a language. Some rules reflect fundamental constraints on speech segments in words, the phonotactic constraints. Thus English does not allow the phoneme sequence /rs/ at the beginning of a word, so there is no initial *rs* spelling for English words. Others are scribal constraints that have developed historically as a result of writing conventions of different sorts (Venezky, 1970). For example, tripled letters do not occur and double letters cannot begin a word. Still others reflect morphological structures, e.g., *ing* is always the spelling of the English present participle and will rarely terminate any word of more than one syllable except as an inflection (exceptions: *herring, awning*).

Notice that these orthographic rules are not rules of grapheme-phoneme correspondence, although the latter are closely related to grapheme constraints, especially morphological ones and phonotactic ones. Again, much has been made about the "irregularities" of English and the hopelessness of a simple correspondence system. However, the rules have to be understood as generalizations that achieve data reduction. There are only a few different phoneme correspondences that are not predictable on the basis of context-sensitive rules. For example, to say that *th* can have several different pronunciations is to miss the point that this pronunciation variability is quite systematic.[9] There are many graphic signals to phonetic values that are reasonably general as well as simple. The rule that a final *e* in a one-syllable word causes vowel tenseness (*mate* vs. *mat*) is one. English has a simple orthography compared with some languages. Its system has been called a "deep orthography" because its spelling tends to reflect underlying morphological structures rather than invariant phonemic correspondences. Thus the spelling of *nation* and *national* preserves their underlying lexical identity rather than signaling their differing individual pronunciations (compare their first syllables). This makes the orthography not less systematic than a shallow orthography (such as for Serbo-Croatian) but merely different in its system. (See Katz & Feldman, 1981.)

The purpose of this digression on rules of orthography is to suggest that such constraints can play a role in lexical access. The role of spelling patterns has been realized for some time (Gibson & Levin, 1975). Even with a description of a rule system, the functional unit in lexical access could be spelling patterns. That is, one could describe the memory of the skilled reader as having a large number of stored patterns. Thus, the rule that *ch* is an allowable string but *hc* is not is represented by there being a *ch* pattern but no *hc* pattern. There will also be a *ct* pattern but it will not be useful for word-initial inputs.

What this amounts to is a level of nonword multiletter representation in memory. In such a model the frequency of a pattern would be indicated by its resting level of activation. Its possibilities for position would be indicated by its embedding in patterns with variables, e.g., *TRx* would indicate that some graphemes must follow *TR*, and *xRT* would indicate that some graphemes must precede *RT*. It is not clear yet whether an independent level of multiletter strings needs to be added to the simpler model that represents only single letters and whole words. However, there is some evidence that sequences of two and three letters, i.e., bigrams and trigrams, are important in word recognition (Massaro, Venezky, & Taylor, 1979).[10] More important for individual differences is the possibility that skilled readers may be especially able to use bigram and trigram sequences in processing letter strings (Massaro & Taylor, 1980).

If rule-based orthographic knowledge is represented directly, for example, by stored bigram and trigram patterns, then we have at least three levels at which lexical access can produce individual differences: letter level, orthographic structure level, and word level. The processing differences that might be affected by these levels are essentially the activation of words in memory. How such processes can make a difference to reading comprehension is considered in Chapter 6. Here we simply ask what kinds of individual differences are found in lexical processing. First we have to consider the implications of our expanded view, namely, that lexical access may include a level of information between single letters and whole words.

Decoding

"Decoding" has been used to refer to different things in reading research, and we need to be clear about its meaning. We have adopted an activation model of lexical access. It is intended to describe the process by which a word is recognized in normal reading. It may, as we have suggested, include a level of representation between single letters and whole words. But none of this is necessarily decoding.

  Decoding is the transformation of a string of letters into a phonetic code. To be clear, there are two things it does not mean. It does not mean the use of a rule-based mechanism. Thus both the more regular word *stop* and the "irregular" word *show* (compare *how, now, cow*) are decoded into their corresponding speech forms. It is quite possible that these cases may involve different underlying mechanisms, as Baron (1977) and others have suggested. But this is not clear (see Glushko, 1981) and not necessarily critical for many purposes. Second, decoding does not mean saying a word aloud. It does not apply just to words but to any codable (phonetically transformable) letter string. If a single mechanism turns out to be sufficient for explaining a phonetic response to both words and nonwords (Glushko, 1981), this will be very important for understanding decoding. Finally, saying a word aloud is a measure of decoding; it is not the process of decoding. Alternative measures are possible.

  There is an implication of this definition of decoding for reading ability: Aspects of lexical access that include (1) graphemic sequences and (2) phonemic sequences below the word level can be sources of individual differences. Thus when ability differences in decoding are referred to, the implication is that they may be important for lexical access differences but not identical to them.

Ability Differences in Lexical Access and Decoding

Do readers of high ability have processes of decoding and lexical access that are more accurate or faster executing? The answer seems to be yes both for children and for adults.

  Children in the elementary grades who differ in comprehension ability also differ in decoding and lexical access. There are two consistent types of difference. High-ability readers are more accurate at decoding and they are faster at it. Their advantage in accuracy may be restricted to words low in frequency and to pseudowords, i.e., letter strings such as *chame* that conform to orthographic rules and are pronounceable. Their advantage in decoding speed is general, but it increases for uncommon words and pseudowords. It also increases for longer words.

The basic result in this area was obtained by Perfetti and Hogaboam (1975). They displayed printed words of various kinds to third- and fifth-grade subjects who were either high or low in reading comprehension. Decoding latency was measured by the time between word presentation and the onset of the subject's vocalization. Decoding latency was shorter for skilled subjects than for less skilled subjects. The difference between groups was smaller for highly familiar words than for less common words. Perfetti and Hogaboam also gave the subjects a vocabulary test following the decoding phase of the experiment. They then considered this question: Was there a relationship between knowing the meaning of a word, as measured by the vocabulary test, and speed of decoding? The answer was "no" for the skilled group and "yes" for the less skilled group. Among low-ability readers, decoding was faster for words whose meanings were known to the subject. For high-ability readers, this relationship was absent. Thus for low-ability readers, but not high-ability readers, decoding seems to be dependent on meaning.

One implication of this research is that individual word decoding abilities are overestimated when the reader is simply required to read the word correctly. This common school practice does not allow the difficulty of processing to be measured. Of course, differences in decoding accuracy are often found between high- and low-ability readers (see also Calfee, Venezky, & Chapman, 1969; Shankweiler & Liberman, 1972). However, if a given sample of words does not produce differences in decoding accuracy, it will in speed.

These decoding-speed differences between high- and low-ability readers are quite wide ranging. They have been as high as 400 milliseconds (Perfetti, Finger, & Hogaboam, 1978) and sometimes as low as 120 milliseconds (Hogaboam & Perfetti, 1978) for vocalization latency. The magnitude of the difference depends on word frequency and word length. Decoding latency differences are less for high-frequency words than low-frequency words (Perfetti & Hogaboam, 1975), greater for two-syllable words and especially for tri-syllables even when frequency is controlled (Hogaboam & Perfetti, 1978; Perfetti, Finger, & Hogaboam, 1978).

The general state of affairs can be summarized as follows: Differences between ability groups in decoding latency increase as a function of word difficulty. This relationship is shown in Figure 5-1, which is based on the data from several studies of this series of experiments, e.g., Perfetti and Hogaboam (1975); Hogaboam and Perfetti (1978); Perfetti, Finger, and

Hogaboam (1978); Perfetti, Goldman, and Hogaboam (1979); Perfetti and Roth (1981). Experimental variables such as real word vs. pseudoword, number of syllables, and frequency have been ignored. Each point is simply a cell mean from some condition of some experiment.

As Figure 5-1 shows, one can predict differences in decoding latencies as a function of overall means. The correlation of .92 means that as a letter string gets more difficult, for whatever reason, ability differences in speed of decoding that word increase.[11]

Decoding and name retrieval

When a subject reacts to a printed word by reading it aloud as quickly as possible, there is a general process of name retrieval involved. *Name retrieval* is the process of retrieving and producing the name of a symbol from memory. It is not just a part of decoding. It is involved when a familiar face is seen and a person's name is produced. Thus a question to ask is whether differences in decoding latency are differences in general name retrieval or in retrieval of names of linguistic inputs.

Perfetti, Finger, and Hogaboam (1978) examined this issue by having third-grade subjects name a variety of stimuli: colors, digits, pictures, and words. Figure 5-2 shows the results. High-ability subjects were significantly faster than low-ability subjects only for word stimuli. There was some nonsignificant difference for digits, and color-naming speed was completely unrelated to reading ability.

The digit-naming conditions of this study are useful for understanding any general name retrieval factor. There were two conditions of digit naming. In one condition, the subject knew that only one two-digit number could occur for a block of trials ("small set" condition of Figure 5-2). Thus there was no name retrieval at all, only preparation of a speech motor program. There were no reader-ability differences and no effect of number of syllables. This contrasts with syllable effects obtained for words and even for digits that were not predictable ("larger set" condition of Figure 5-2). Syllable effects and reader-ability effects seem to depend on operations from memory, in particular an *inactive* memory.

The conclusion from the experiment just described is that general name retrieval speed is not the major source of decoding speed differences among children. However, such differences in name retrieval may be present in comparisons of more disabled readers (Denckla & Rudel, 1976a, b; Spring & Capps, 1974) or in situations which require repeated responding rather than single-item retrieval. It should be kept in mind that our discussion is restricted here to the range of reading ability found in a normal classroom.

Another important result of the Perfetti et al. (1978) experiment concerns the differences between time to name a word from an open set and

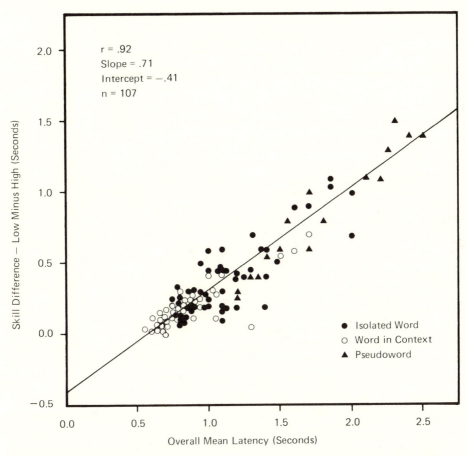

Figure 5-1. Differences between high- and low-ability readers in decoding latency depend on the mean decoding latencies for all subjects. From different experiments using vocalization latency to words or pseudowords as the dependent measure.

from a closed set (Figure 5-2, right panel). A closed set is exemplified by the names of the four seasons or the twelve months. An open set is *dogs* or *proper names*. Thus an item from an open set is less predictable than one from a closed set. High-ability readers were faster overall than low-ability readers, but their advantage was especially large for an open set that had a large number (12) of items. This points to the importance of inactive memory processing in the skilled reader's advantage. That is, when an item is unpredictable and therefore has little activation prior to its presentation, skilled readers are faster.

Decoding without naming
Vocalization latency is not the only measure of decoding that produces

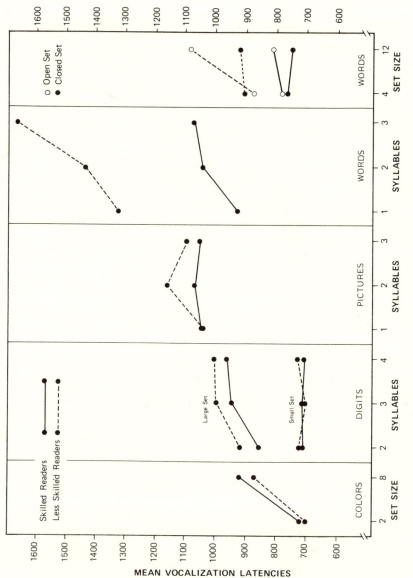

Figure 5-2. Vocalization latencies of skilled and less skilled readers for colors, digits, pictures, and words. The words of panel 4 are the names of the pictures of panel 3. Reprinted from Perfetti, Finger, and Hogaboam (1978).

ability differences at the word and nonword level. Here are other decoding tasks that have been used in our Pittsburgh research.

Visual matching. In an experiment described in Hogaboam and Perfetti (1978), subjects were presented with pairs of words or pseudowords. When the letter strings were the same, the subject responded by pressing a button. The two strings were separated on the viewing screen, thus requiring at least one eye movement. Skilled fourth-grade readers were faster at the matching task than less skilled readers, but the differences were smaller and less reliable compared with vocalization latency data on the same subjects.

Spoken-written word matching. In an experiment described in Perfetti and Lesgold (1979), third- and fifth-grade subjects had to decide whether a visually presented word matched the word just spoken to them. Again, a manual response was required. Differences between high- and low-ability readers in this measure of decoding were smaller and less reliable than those obtained for vocalization latency (about 50 milliseconds, compared with an average of 200 milliseconds in vocalization latency). This difference was larger than for a comparable decision task in which a picture was presented instead of a word, but not significant. The same experiment, however, found a larger and more reliable difference (150 milliseconds) when the subject had to decide whether the word or picture belonged to a specified semantic category.

Lexical decisions. In experiments previously unreported, subjects in the fourth grade performed lexical decisions on words, pseudowords, and nonwords. Pseudowords are pronounceable and regular, while nonwords are not. Skilled readers showed a significant advantage in speed for all classes of items. The mean decision times were as follows for nonwords, pseudowords, and words, respectively: Less skilled readers, 1.70 seconds, 2.10 seconds, 1.95 seconds; skilled readers, 1.50 seconds, 1.60 seconds, 1.55 seconds.

The general conclusion from all these experiments, which were carried out on the same subject population, is this: Speed-of-decoding differences are general over a range of tasks. However, there are different processes tapped by different tasks, and these differences are significant. A task that involves the production of a word name without context leads to consistently large and reliable ability differences. Tasks that can have a large visual component and tasks that provide the subject with part of the information needed produce smaller and less reliable differences. Thus the visual matching task, which unfortunately measured only matching time and not mismatch time, can require a match based on an essentially visual basis, a synthesized visual code. And the spoken-written word-matching task provided the subject with the spoken word first, thus causing some activation of the word the subject was about to see.

Other Research. This discussion of lexical factors in reading has been very selective. It has referred to research that was carried out on comparable populations and that provided part of the basis of the verbal efficiency theory. There is, in fact, a large body of evidence concerning ability differences in lexical processes. (See, for example, Vellutino, 1979).

SUMMARY

This chapter has discussed ability differences in reading comprehension by referring to comprehension and lexical processes previously described. Reading comprehension is the ability to be explained, and there are four general sources of ability differences in comprehension: (1) the availability and application of schemata in constructing a text model; (2) the encoding of text propositions in working memory; (3) knowledge of word meanings needed for a text; (4) lexical access processes required for reading the words of a text.

Our analysis of schemata-based factors in comprehension leads to the conclusion that a reader's knowledge is critical in the construction of a text model. We focused especially on research that shows that domain-specific knowledge affects a comprehender's representation of a text. The main differences between those who have knowledge of the domain and those who do not center on the higher level structures and how lower level actions relate to them. However, the importance of schemata for comprehension does not mean that a schema explanation of general reading ability is promising. In fact, while it accounts readily for individual differences in comprehension of texts that depend on specific knowledge, its application to an understanding of *general* reading ability is limited. A companion emphasis on the metacognitive processes by which a reader controls his comprehension may account for some general reading-ability differences, but more evidence is needed.

With respect to propositional encoding in short-term memory, the evidence is clear. High-ability readers remember more of what they have just heard or just read than do low-ability readers. This means that differences in comprehension at high levels, e.g., construction of a text model, must include the local processes operating in sentences and propositions. The ability to hold in memory verbatim text information will largely determine the ability to recall a text later on.

A second local processing factor is the ability to encode word meanings. High-ability readers know more word meanings than low-ability readers. They appear to be vaguely familiar with more words and to have a more precise understanding of words. These factors enable them to encounter texts with less risk of comprehension failure. On the other hand, it is not clear that a word meaning explanation is general enough. To the extent that reading ability differences are observed for texts which contain only familiar words, additional explanations are required.

The final factor considered was the lexical processing level. It was suggested that a model of lexical access should include a representation of the facts of English orthography. Readers may differ in the extent of their orthographic knowledge, and a model of nonword multilevel representation in memory seems helpful in explaining the decoding of letter strings that are not words. Decoding, defined as the transformation of a printed letter string into a phonetic code, is a process that sharply differentiates children who differ in reading ability. This seems especially true for measures of decoding that require naming and for situations that require the name to be retrieved from an inactive memory.

NOTES

1. The words "skill" and "ability" are used interchangeably throughout. "Ability" is not used to imply skills that are resistant to learning, i.e., it is not used to connote "innate ability."
2. Actually, the dominant interpretation of this passage is that it is a vague and oddly phrased text that could only have been written for an experiment. However, the Rocky passage is better than most in this respect. For evidence that vagueness of reference is not a necessary feature to demonstrate the role of knowledge, see Figure 3-2.
3. Spilich et al. (1979) tested a model that assumes a separate memory system for macrostructure propositions, the higher level information of a text (Kintsch & van Dijk, 1978). They found such a model to provide a better fit than the model of Kintsch and van Dijk, which assumes that macrostructures are constructed from a short-term memory containing only propositions (micropropositions).
4. For example, there is evidence concerning expert-novice differences in chess (Chase & Simon, 1973) and physics problem solving (Larkin, McDermott, Simon, & Simon, 1980). Some of these differences clearly involve *representation* differences among levels of some ability that are already high. In the case of physics, problem structures are strikingly different in expert and novice. In the chess case, one is tempted to say, not that the structures are qualitatively very different, but that the expert's is "richer" in some sense having to do with thousands of hours of additional experience with the same structure.
5. When he returned from his travels abroad, Odysseus killed all his rivals and all his female house servants who had had anything to do with them. Presumably he and his faithful wife then lived happily ever after.

6. In reading a long, difficult sentence, memory for the first part of a sentence can be quite poor. This is what Goldman et al. (1980; Experiment 2) found for less skilled third graders.

7. Some of these issues turn on the relationship between IQ and reading. For example, digit span is more likely to be related to reading ability when IQ is not controlled (Belmont & Birch, 1966).

8. A similar distinction is captured by *breadth* vs. *precision* (Cronbach, 1942) and *range* vs. *precision* (Kirkpatrick & Cureton, 1949).

9. F. Smith (1973) despaired of finding any order in the chaos of English. He claimed that *th* was "completely unpredictable" and contrasted *this, than, that* with *think, thank, thatch,* and *thong.* The big difference, of course, is simply the phonetic feature of voicing, a rather systematic feature difference. In fact, whether the voiced or voiceless *th* occurs in a word-initial position is not unpredictable. The voiced *th* tends to be restricted to simple consonant-vowel-consonant syllables. In any case, it is difficult to imagine this difference having any effect on a reader.

10. Note that the whole-word model (Rumelhart & McClelland, 1981) can represent the importance of subword units such as bigrams and trigrams. This is done indirectly by the occurrence of letter strings as part of words. For example, *ch* occurs at the beginning of many words and *ct* never occurs. The reason for representing bigrams and trigrams ordinarily is to represent the advantage that permissible letter strings, pseudowords, have over impermissible ones in recognition and search tasks. Thus a pseudoword such as *chame* is processed more readily than *ctame.* This can be accounted for either by representing knowledge about orthographic rules, or by representing only whole words. For the latter case, Glushko (1981) demonstrates that processing a pseudoword like *chame* will depend on its letters receiving more activation from real words, e.g., *shame, champ, chime.* By such an account, orthographic rules are not required. Such parsimony perhaps should not be disturbed. However, the model of Rumelhart and McClelland (1981) and its application by Glushko (1981) has been restricted to four-letter words. It remains to be shown that intermediate levels of representation are not needed to recognize, say, six-letter words and pseudowords. Meanwhile, such a level of subword representation does capture directly some of the individual differences in use of orthographic constraints.

11. There are various interpretations of the correlation. For example, the process that takes a long time to execute is one that is prone to error. Thus "more difficult" words will have higher error rates than less difficult words. On the other hand, one might say that such a correlation between difference scores and means is just a statistical fact that causes a problem for the analyses of variance that were done in the separate experiments. That is, there may have been a correlation between means and variances. In fact there usually are in such cases, courtesy of the real world. It is my conclusion that nothing theoretically important hinges on such problems.

CHAPTER SIX

Verbal Efficiency Theory

This chapter continues the discussion of individual differences that was initiated in Chapter 5. Verbal efficiency theory is a theoretical framework to explain such individual differences. What needs to be explained are the general facts of individual differences in reading comprehension ability: that is, that some people read rapidly with comprehension and others read more slowly and/or with less comprehension. We have already seen one important consideration that must now be taken seriously: There cannot be a theory of *general* reading ability that depends on *specific* knowledge. The alternative to this conclusion is that the concept of general ability is an illusion.

If there is such a thing as general reading ability, then it must explain the salient facts of individual differences in reading comprehension, namely, that differences in text comprehension, as evidenced in text recall or text question answering, arise in part at local processing levels. Low-ability readers do not simply fail to retrieve information necessary to demonstrate comprehension. The evidence indicates that low-ability readers are already "missing" text information a few seconds after they process it. This is the force of the Perfetti and Goldman experiments described in Chapter 5. A second salient fact is that high-ability readers outperform low-ability readers in decoding single words, especially where production of the word is required and where the word has no prior activation. A third salient fact is that many differences between high- and low-ability readers are present for listening as well as reading.

Verbal efficiency theory is a model of skill in reading that focuses on these facts. It explains them essentially by reference to the efficiency of linguistic processing provided by knowledge-rich and quickly executing lower level processes. The general features of this theory are described first, followed by a more specific account, in which critical questions are raised.

GENERAL FEATURES OF VERBAL EFFICIENCY

In its general form, the theory of verbal efficiency is that individual differences in reading comprehension are produced by individual differences in the efficient operation of local processes. The local processes are those by which temporary representations of a text are established. In terms of the components of reading discussed in previous chapters, verbal efficiency assumes that construction of an effective text model depends critically on the assembly and integration of propositions in working memory. Therefore any processing factor that contributes to the encoding of these propositions will affect overall reading comprehension. Individual differences in reading comprehension are produced by these local processing factors. This is not to say that comprehension differences are not also produced by schema-related processes. Other things being equal, differences in appropriate schema activation will produce large differences in reading comprehension. One of the important empirical questions is the extent to which other things, i.e., verbal efficiency, tend to be equal among individuals.

General Assumptions

The theory entails certain assumptions widely validated in research on cognitive processes. For one, it is assumed that some mental processes operate under limitations imposed by system constraints. This is the assumption of a limited-capacity working memory system (Baddeley & Hitch, 1974). It is not critical whether this limited-capacity system is conceptualized as part of a structurally separate short-term memory (Atkinson & Shiffrin, 1968) or as a functionally distinctive part of a general permanent memory (Anderson, 1976). However, because the latter conception lends itself to describing activation processes, we prefer this description: *Working memory* is the limited-capacity processing system that is constrained by the number of memory elements that can be simultaneously activated. These elements include not only permanent memory nodes such as words but also the temporarily constructed links among nodes. That is, working memory is used for the comprehension of sentences. It stores the results of partly processed sentences, for example, the first phrase or clause, and it groups words into tentative structures as they are encountered.

A related assumption concerns *attention.* Theories of attention have tried to address some of the same phenomena described by a limited-capacity working memory. The assumption that some processes can occur without attention while others require attention is the equivalent of saying that limited resources are applied only to the attention-demanding processes (Kahneman, 1973; Shiffrin & Schneider, 1977; Posner & Snyder, 1975). In addition, attention concepts provide two significant considerations for verbal efficiency. One is that the extent to which a process requires attention is not fixed by the processing environment. It is affected, within limits, by the amount of overlearning (practice) given to the process (LaBerge & Samuels, 1974; Schneider & Shiffrin, 1977). This does not necessarily mean that processes can become completely attention-free with practice. But we do assume that the amount of attention required is modifiable by processing experience. This possibility has been demonstrated for the simple reading processes of letter recognition by LaBerge and Samuels (1974).

The second attention assumption is *resource allocation*. In a given processing environment, the demands on attention of competing elements are met in part by the individual's control procedures. These procedures in effect cause more processing resources to be allocated to one process than another (Kahneman, 1973; Norman, 1976).

These assumptions are general in that they apply to cognitive processing generally. They are reflections of basic constraints on human information processing and control procedures adapted to these constraints. Even allowing for demonstrations that skilled performance can overcome capacity limitations under some conditions (Neisser, 1976), these assumptions are generally secure. They provide the general basis for the core proposal of verbal efficiency theory.

Central Features of Verbal Efficiency Theory

The core proposal of verbal efficiency theory is that the outcome of reading, i.e., comprehension of what is read, is limited by the efficient operation of local processes. There are some specific components to verbal efficiency in this sense. These include the definition of efficiency and a consideration of the range of efficiency for different processes. The latter question is essentially what processes can be made efficient. We suggest that this is partly a matter of how processes contribute to the reader's text work. The distribution of resources for the text work is the basis of individual differences. The distribution of resources is not totally under individual strategic control, but is dependent on the reader's ability in low-level processes.

Verbal efficiency
Verbal efficiency is a concept of product and cost. The product is a degree of reading process outcome and the cost is the processing resources required to achieve the outcome. The outcome of a reading process can be the identification of a word, a semantic decision about a word or phrase,

or the comprehension of any text unit. It can, in fact, be any reading outcome. It is called *verbal* efficiency because the outcome of a reading process is verbal. More important, however, is the claim that what must be
efficient are processes that are in large part verbal; that is, they prominently include semantic, orthographic, and phonetic components.

Verbal efficiency is the quality of a verbal processing outcome relative
to its cost to processing resources.

Processes that vary in efficiency

Any reading performance can be more or less efficient in the sense of
this definition. Thus we can say that an entire text has been efficiently
processed. However we are clearly referring in such a case to unspecified
processes that have contributed to the overall achievement. Was it the
efficient activation of relevant schemata, the efficient access of word
meanings, or the efficient integration of propositions? In principle, any
of these processes can contribute to overall efficiency.

Schema activation. The very use of the term *activation* implies a potentially high-efficiency process. Efficient schema activation is "text-driven"
in part. That is, a familiar text form and a familiar topic "automatically"
activate appropriate schemata.[1] On the other hand, there are many cases
of inefficient schema activation. In such cases, the processes may include several search and comparison processes. The text continues to activate candidate schemata but subsequent text samples fail to fit. The situation in vague texts can be described as the *weak activation of multiple
schemata.* However, inefficient processes occur as the reader compares
these schemata with text samples. This is process-costly reading work.

It is an important practical implication of verbal efficiency that efficient schema activation is not only possible but reasonably likely for
mundane texts and reasonably experienced readers. As for individual
differences in schema activation, they are less likely as general characteristics (see Chapter 3). That is, individuals, irrespective of reading ability, can ordinarily be expected to have schemata activated automatically
provided two conditions are met: (1) They have the necessary knowledge
in the first place; (2) Their low-level text processes are sufficient to trigger their activation. It is an important consequence of verbal efficiency
theory that *inefficient local processes will often cause schema activation
to misfire.*

Propositional encoding. The efficiency of proposition encoding is variable because it involves multiple lexical processes. Several words may
be needed for a working-memory representation that can be linked up
with the reader's current representation. It is possible that propositional

encoding could in some sense be "automatic," but it is probably better to say it this way: The encoding of text propositions can range from fairly high to extremely low efficiency. The integration of encoded propositions into underlying text structures, for example, establishing causal links, may be especially resource costly.

As for individual differences, the limiting factor is working-memory capacity. However, the assumption is that this is a *functional* limitation that is only partly due to structural limitations. That is, two individuals may have equivalent short-term memory capacity (whatever that might mean exactly) and have different functional memory limitations. The individual with the larger functional capacity will be a more efficient processor of text.

Lexical access. In reading, lexical access processes range from "automatic" to resource costly. However, these processes are the most likely candidates for becoming uniformly high in efficiency. This assumption can extend also to encoding the semantic information stored with a word. An activation mechanism potentially enables semantic information required by context to be triggered "automatically." Also, importantly, the speech code for a word can be automatically activated, enabling a reference-secured code for memory.

The potential for high-efficiency lexical access is important for working memory. To the extent that lexical access is resource efficient, the encoding of propositions in working memory can be achieved more efficiently. This means that individual differences in working memory can arise either because of factors characteristic of propositional encoding, e.g., ordering elements and encoding predicate relations, or they can arise from inefficient lexical processes that compete for the same resources.

The Text Work

The general account of verbal efficiency is that each of the processes of reading can have a range of efficiencies. However, in some absolute terms, the limits on efficiency are different for different processes. Thus, a proposition-encoding process of maximum efficiency will still use more resources than a maximally efficient lexical access process.

Individual differences are matters of efficiencies for these subprocesses. The reader of high ability confronting an ordinary text has high-efficiency lexical access processes. There may be, in some sense, an automatic access to words as a result of the reader's experience in reading (LaBerge & Samuels, 1974). In addition, the text and the reader's knowledge interact to cause the activation of higher order knowledge structures. These structures are activated mainly by properties of the text, but the reader must have them in order for activation to occur. In the ideal

case, these lexical and schematic processes can take place with little resource expenditure. That leaves the resources for the work that needs them:
(1) the encoding of propositions, especially the integrating of propositions within and across sentences; (2) some of the inference processes
that are not automatic, i.e., the kind of memory search that is needed when
a text contains a gap; also perhaps the kind of reinstatement needed for
recently backgrounded information (Lesgold et al., 1979); (3) the inter-
pretative, inferential, and critical comprehension of a text that goes beyond the text itself—linking a mental text model to knowledge structures
that require explicit (and perhaps novel) linking even if their original activation was "automatic."

These three text-processing activities are closely related in time and
fundamentally nondistinct. In fact, they take place within the limited-
capacity working memory. They add up to the *text work* of the reader.
The specific claim of verbal efficiency is that the text work is made easier to the extent that those processes which *can* be at high efficiency *are*
at high efficiency. Such processes include especially lexical access and
perhaps *elementary* propositional encoding. This elementary proposition encoding is distinct from the kinds of integrative and inferential
proposition work that are higher in resource cost. The elementary encoding essentially consists of *assembling* a single proposition from only a
few words, typically two or three.

Figure 6-1 represents some of the text work that might go on in reading
the simple doctor's visit story (see Chapter 3). Represented is some of the
text work that occurs just during the reading of the second sentence: *The
room was warm and stuffy, so they opened the window.* A time line indicates plausible times for four different kinds of text work: *lexical access, proposition assembly, proposition integration,* and *text modeling.*
The last provides the basis for the interpretive, inferential, and critical
comprehension referred to above. The processes are represented as overlapping in time. (As before, we assume that lexical access can be influenced by prior activation; this is not a strictly bottom-up model.) For
convenience, it is assumed that lexical access occurs for eight of the eleven
words in the sentence and that the reader must assemble six propositions, including one that is integrative. An integrative proposition is one
that requires some links to be established with prior propositions.[2] Finally, there is an estimation of the resource cost of the processes. Above-
average amounts of text work are indicated by an asterisk; still greater
amounts are indicated by two asterisks. Assuming a skilled reader, the
suggestion in Figure 6-1 is that the most costly single piece of text work
is constructing the *because* proposition. It links two assembled propositions that are separated by several words.

Notice that it is assumed that incomplete assembly processes occur.

Time →	Access	Assembly	Integration	Text Modeling
	The room	(1) exists (room)		$\begin{bmatrix} \textit{wait (they, doctor, } \phi) \\ \phi = \text{room} \end{bmatrix}$
	was	was room		
	warm	(2) warm (room)		
	and stuffy	(3) stuffy (room)		
	so	so?		
	they	so they	*they = Joe & infant daughter	
	opened	open they		
	the window.	(5) open (they, window)	**because (5, 2 & 3)	? complication for visit
		(4) exists (window)		? significance in event structure
Total Resource Cost	A	P	I	M

Figure 6-1. A representation of hypothetical text work for the sentence *The room was warm and stuffy, so they opened the window* when it is the second sentence of the doctor's visit story. Represented are lexical access, assembly and integration of propositions, and text modeling. Resources needed for the text work are represented at three levels. Unmarked processes are low and roughly equal. Asterisks indicate processes that may require more work.

105

These incomplete processes occur for at least four words in this sentence (the unnumbered ones are incomplete). Thus, *so* triggers an incomplete proposition and *they* is attached to it until finally *opened* is encoded to fill in the proposition. Presumably, delay in final proposition assembly can be a significant source of text work. (Imagine, for example, an embedded clause between a noun and its verb.) This is because working memory must try to hold onto partially assembled propositions until they can be completed.[3]

The last row of Figure 6-1 is a hypothetical distribution of the total cost of the text work. The assumption is that for any processing outcome, there has been a distribution of effort among lexical access, assembly of propositions, integration of propositions, and text modeling. These are the parameters A (access), P (assembly of propositions), I (integration), and M (text modeling) of Figure 6-1. These parameters vary widely from one sentence to another (or some other work unit). They can also vary from one individual to another. If the total text effort is T, then $A + P + I + M = T$.

Of general interest, over large sections of text, are the relative magnitudes, A/T, P/T, I/T, M/T. In cases of easy texts and skilled readers, I/T should be high relative to the others. And A/T and M/T should be low. The normal high-efficiency resource allocation may produce $I/T > P/T > A/T = M/T$. To put it another way, in arriving at a text representation over, say, two or three sentences, schema activation and text-driven inferences (M) and lexical activation (A) may be relatively easy. Integration of propositions (I) will be relatively difficult.

All this should be taken only as a generally plausible description because we can offer no serious estimations of these parameters in any given situation. However, to continue with a bit of speculation, the following additional considerations become interesting. Suppose a text is made more difficult by substituting less frequent words for familiar words, but otherwise not changing anything. For the same two or three sentences, T (the total difficulty) will increase. Furthermore, every component of T will increase, not just A. Schema activation will be more effortful, insofar as it depends on easy word recognition, and it may now require active memory search. Thus M will increase. So too will P and I. Propositions are more difficult to assemble and to integrate partly because the words are more difficult to access. However the ratios of these components for total effort will also change. A, P, I, and M will all increase, but A/T may well increase more than P/T or M/T. This follows simply from an additive assumption: make the words more difficult by some constant factor and all components of text processing that depend on word encoding are increased by this factor. The ratios of less resource-costly processes will necessarily increase more.

A text work example

The text work analysis rests on only weak assumptions regarding the relationship among the text work components. It assumes a weakly interactive relationship in which the "lower level" processes must reach some minimal level before the "higher level" processes are *begun*. It allows the completion of a higher level process to occur at about the same time as completion of a lower level process.

This is represented in Figure 6-2, which shows four text events occurring over a matter of half a second or so. The reader has already read "Joe and his infant daughter were waiting for the doctor to return from lunch. The room was warm and stuffy, so they . . ." Shown here is the completion of four events that follow rapidly at just this moment: (1) The word *opened* is accessed. (2) The phrase *the window* is accessed with a fixation occurring just on *window*. (3) The proposition *open (they, window)* is completed and connected to the already stored proposition about Joe and his infant daughter (= *they*). (*They* has already been encoded, so the proposition assembly process has already begun before *open* and *window* are accessed.) (4) The *because* proposition linking *window opening* to *warm* and *stuffy* is assembled and integrated. It has begun with the word *so*. What is shown in Figure 6-2 is the completion of the process.

The assumption that some text work is begun before other text work is completed is reflected in Figure 6-2 in that the last two text work functions are above zero during the execution of the first two functions. That is, while the lexical access of *window* is occurring, there is the beginning of a proposition assembly process for *open (they, window)*. It has been triggered by the word *they*, pushed forward by the word *open*, and completed by the semantic encoding of *window*. A similar description is possible for the *because* proposition. It was triggered by *so* and completed by a memory search or reinstatement following assembly of the window-opening proposition.

In summary, there are two important assumptions about text work that are reflected in Figure 6-2. The text work of proposition assembly and proposition integration overlap in time with lexical access. This holds as well for text modeling, which is not shown in Figure 6-2. However, such work does not approach completion until lexical access and case assignment are completed. By "case assignment," we refer to encoding the critical relationships between elements of a proposition. It is a major component of what we have been calling *propositional assembly*. In the window-opening example, it means the encoding of *window* as the recipient of the opening action and *they* as the agent of this opening.[4]

Overall, the analysis of text work leads to two general conclusions. For one, the various components of text work do not have to be thought of

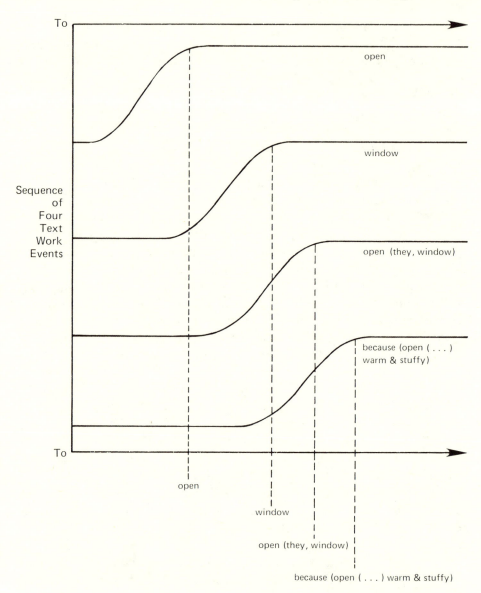

Figure 6-2. Overlapping text work for four text work events. The reader is encountering part of the second sentence of the doctor's visit story. Each text work event is represented as a rising nonlinear function that is "completed" at asymptote. Input from one text event is used by the next. However, there can be a beginning of a given text work prior to the completion of a logically prior one.

as a series of processing stages. They may be better conceived as components, inseparable for some purposes and overlapping for other purposes. The other conclusion is that the components of text work vary in their share of effort from text to text and person to person. However, if we think of an ideal reader and an ideal text, it seems clear that some components, especially the lexical access and the schema components of text modeling, are potentially less resource costly than the propositional components of assembly and integration. While this text work analysis is hypothetical and has not been verified in detail, it is a psychologically plausible account that, in part, can be experimentally tested.

Text work and reading ability

The application of the text work analysis to reading ability assumes that the text work components vary with texts and individuals. The text work required by a given reader i for a given text is a function of the idealized work components of the text averaged over individuals (T_j) and the idealized work resources of the individual (T_i) averaged over texts, i.e., $T_{ij} = f(T_i, T_j)$. This is a straightforward and very weak assumption. It says nothing about f, the function relating ideal individual ability and ideal text difficulty.

However, it does have some interesting implications. For example, assume a series of texts ordered according to increasing difficulty and a series of individuals ordered according to decreasing ability:

$$\text{Texts} \quad G<H<I<J<K<L$$
$$\text{Individuals} \quad g>h>i>j>k>l$$

This means that there exist many individuals for whom text J is more easily read than text K. Also it means that there exist many texts such that reader i reads them more easily than reader j. This is the essence of a concept of reading ability. It is just the sort of thing that text-specific knowledge does not yield. It also implies that there exist two different texts, of difficulty J and K, such that individual i reads text J as well as individual h reads text K. (For text J, $h>i$; and for text K, $h>i$.) The number of such texts depends on the limits of the function relating text difficulty to individual ability, but there must exist at least one such text for any two individuals within a narrow range of ability.

This scheme provides a general description of ability differences and text difficulty. It depends on the assumption that text difficulty should be described by reference to known processing factors (Kintsch & Vipond, 1979). It also probably depends on measuring text difficulty in terms of comprehension product per unit effort (or time). Within limits, it is possible for two individuals to show equal comprehension when time is uncontrolled but not when time is kept short. In fact, this trade-off may not apply commonly to the range of ability differences between children that we have considered. For many low-ability children, it is unlikely that mere increases in time for reading will produce comprehension equivalent to that of a high-ability reader.

The assessment of individual ability differences is made largely in terms of a global measure of the product of reading. We wish to understand why two individuals can read the same texts and, with only weak limits on reading time, differ in their ability to answer questions about what they have read (even with the texts still available for inspection). Verbal efficiency explains this by assuming that the individuals differ in the amount of text comprehension they achieve per unit of text work. It further assumes that the key to these differences in resource cost is the potential of reading components for achieving low resource cost. Lexical processes are one component with a low cost potential and these are likely candidates for fundamental ability differences. Furthermore, an increase in their cost is potentially devastating for later occurring processes that depend on completed lexical processes. These later processes include schema activation. Thus verbal efficiency theory claims that unobserved ability differences in lexical access produce observable differences in the use of schemata during reading.

Text work and readability

For adult readers, it is sometimes difficult to see how such low-level processes can be so important for comprehension. For such cases, it is useful to try to read something that simulates some of the problems a low-ability reader has. This change in comprehension for a given reader with a change in text is part of readability.

> *The ephemerella transformations, unlike those of caddis (sedge), include four distinct morphological structures, larval, nymphal, imago, and spinner. The first two are protracted subaquatic stages, while the third and fourth are aerial and truly ephemeral. The angler's palpitations on a June evening usually result from rises to these latter transformations.*

This text is not really technical, but its three sentences are not well comprehended by some readers without effort. Appropriate schema activation is important and can help compensate for the local processing difficulties. So also can a slight rewriting that does not change any of the meaning:

> *The may fly goes through four stages of development (unlike the caddis fly)—larval stage, nymph, imago, and spinner. The first two stages occur underwater and last a long time, while the last two occur in the air and are over quickly. Trout feeding on these last two stages are the usual cause of a fly fisherman's excitement on a June evening.*

The changes made in this second text are not limited to the substitution of more familiar words for less familiar. There are still three sentences, but 61 words instead of 50. Semantically dense words have been semantically expanded: *angler* → *fly fisherman*; *rise* → *trout feeding on*; *protracted* → *last a long time*. The fact that some of these alterations seem to reduce specialized jargon *(rise)* while others reduce general vocabulary unfamiliarity *(ephemerella, protracted)* is probably not relevant. At least one alteration actually seems to increase ambiguity: *distinct morphological structures* → *stages of development*. And at least one eliminates a nonhelpful elaboration in favor of one that promotes referential coherence, *caddis (sedge)* → *caddis fly*. Only one is a single lexical change toward familiarity: *palpitations* → *excitement*. Overall, the effect of these changes is to increase the number of words and the number of propositions—11 additional words and about 8 more propositions. However, the revision does not increase the ratio of propositions to words, but actually decreases it slightly. Thus there is a very slight reduction in the number of propositions that must be constructed per word.[5]

Thus, the character of the revision is one that should make for slightly easier text for a number of different reasons, not a single one. It would be a mistake to suggest that verbal efficiency is related to text difficulty simply at the single-word level, or that options for simplifying text are mainly word choices. For example, sentence (1) can be rewritten as sentence (2).

(1) *The ebony feline dispatched the albino rodent.*
(2) *The black cat killed the white rat.*

This alteration duplicates the original syntax exactly, but it represents a very rare option for text production. Simple, one-for-one lexical choices are not at the root of all text difficulty.

The reason for raising this issue is to make clear that neither text difficulty nor verbal efficiency is to be equated just with single-word processes. In both text difficulty and verbal efficiency the critical feature is the extent to which local text processes are made easier. Difficult words can make proposition assembly and integration more difficult. However, the general principle is that any text feature that increases the effort of propositional encoding affects text difficulty. For reading ability, the corresponding principle is that any local process that increases the effort needed for propositional encoding can produce comprehension failure. This can happen because the reader abandons the difficult job of construction of a coherent text representation or because severe compromises with the text meaning must be made in order to construct any representation at all. The importance of lexical access is that it is a potentially high-efficiency process; when operating efficiently, it can enable the reader to work more on propositional integration even as the text becomes more difficult. Other local processes are critical as well in this respect, a point we will return to later.

A related concept of readability

There are a number of quantitative procedures that are commonly applied to a text to measure its readability (see Klare, 1974/75). However, as Kintsch and Vipond (1979) observed, theories of text processing should be able to contribute some theoretical understanding to these measures. Kintsch and Vipond demonstrate that the intuitive difficulty of some texts is predicted by an analysis based on their model of microstructure processing. In fact, they are able to make five specific suggestions in terms of their processing assumptions. Two fall under the category of proposition assembly: (1) the ratio of propositions to words and (2) the ratio of *different* arguments (nouns and other propositions) to the number of propositions. These first two factors are based on data showing that reading time for a text depends on the density of its propositions (Kintsch & Keenan, 1973) and the density of *new* arguments (Kintsch, Kozminsky, Streby, McKoon, & Keenan, 1975). In addition, Kintsch and Vipond identify three additional factors that go beyond propositional assembly: (3) the number of memory researches required by a failure to find a linking proposition in short-term memory, (4) the number of inferences required by lack of text explicitness, and (5) the number of reorganizations required by the reader's attempt to achieve an optimal organization. Factors 3 and 4 fall under the category of propositional integrations, while the last seems to be an example of text modeling.

What is especially interesting is that those measurable text features can be related to model parameters that determine (1) how many propositions get assembled at once ("input chunk size") and (2) how many propositions can be held in working memory. By adjusting these parameters, Kintsch and Vipond were able to show that a text that is difficult for a reader with low memory capacity (or low "input chunk size") is much easier for a reader with a higher memory capacity (or higher "input chunk size"). Moreover, which of two texts is more difficult could be reversed by changing the relative magnitudes of these two proposition processing parameters.

Thus, our discussion of text work and reading ability is fully compatible with the idea that text difficulty can be assessed by text-processing parameters of the Kintsch & van Dijk (1978) model. However, as it stands, there has been little effort to apply the two individual difference parameters of their model, "input chunk size" (n) and short-term memory capacity (s) to individual differences.

With respect to verbal efficiency, it is especially important to realize that the lexical access *(A)* and the assembly of propositions *(P)* are non-trivial for many readers and for many texts. Neither of these is well captured by the Kintsch and van Dijk (1978) model, which, understandably, is intended to presuppose these processes rather than explain them.[6]

SPECIFIC COMPONENTS OF VERBAL EFFICIENCY

The general account of verbal efficiency links total text work to local processes that vary in efficiency. It assumes that reading ability consists in part, of being able to allocate resources in accordance with the text work demanded by an ideal text. A corollary of this assumption is that reading ability includes the ability to modify resource allocation in response to the variable demands of actual text. This ability is not merely one of allocation strategy, but of what is possible given the demands of the task and the ability of the reader.

There are several possibilities for a more detailed account within this framework. Most prominently, there are two classes of possibilities for the primary component of verbal efficiency: lexical access and propositional encoding (assembly and integration). Lexical access is a process that, in isolation, does not approach the functional limits of working memory, whereas propositional encoding, especially propositional integration, seems to be partly dependent on memory limitations. Thus individual differences in verbal efficiency may stem primarily from lexical access or primarily from memory limitations, or from both. Any of these circumstances could produce observed differences in working-memory capacity. However, by one interpretation these differences are fundamental process limitations, while by the other interpretation they arise because of inefficient lexical access.

Lexical Access Hypotheses

One way to elaborate verbal efficiency theory is to assume that lexical access is the critical feature. Efficient lexical access, rapid and low in resource cost, enables working memory to carry out the propositional text work. Inefficient lexical access, slow and efforful, makes it more difficult for working memory to do this work. This general proposal was made in Perfetti and Hogaboam (1975) and elaborated in Perfetti and Lesgold (1977, 1979), Perfetti (1977), and Lesgold and Perfetti (1978).

The general lexical access hypothesis can have several specific forms. Two of these are that (1) inefficient access is interfering, and (2) inefficient lexical access produces low-quality codes.

Lexical access interference
The specific manner in which comprehension depends on lexical access

can be described as follows. The assembly and integration of propositions are interfered with by a process that competes for processing resources. Inefficient lexical access disrupts the temporary representation
of text in working memory. For an example, refer back to Figure 6-1, which
shows the hypothetical text work for *The room was warm and stuffy, so
they opened the window.* On encountering *opened,* the reader is holding
onto propositions about the room being warm and stuffy which will have
to be linked to the *so* clause. The reader has also begun to assemble the
so proposition and the *they* proposition (see Figure 6-2). With this text
work going on in working memory, the word *opened* is accessed. A rapid
effortless access of *opened* produces the semantic (and phonetic) codes
necessary for two propositions and *window* quickly follows to complete
all the partially assembled propositions. If the reader has trouble with
opened, i.e., if it takes significant processing resources away from holding onto the already assembled propositions and the partly assembled ones,
then memory for the latter are at risk. The same story applies to the access of *window* except that at *window* there is more text work to perform. It completes two propositions.

This particular text is very easy. If we imagine a slightly more difficult
text, it is easy to see how minor problems with a single word can ex-
plode to disrupt the text representation. For example, consider the other
example of this chapter: *The ephemerella transformations, unlike those
of caddis (sedge), include four distinct morphological structures. . . .*

Code quality
The second specific mechanism for the dependency of comprehension
on lexical access focuses on lexical access itself. It is not just that inefficient access interferes with working memory. Rather, the code that results from inefficient lexical access is of low quality. This low quality
can be described as insufficient semantic activation, insufficient phonetic activation, or code asynchrony. Although these descriptions of quality
seem to be different, they depend on the common assumption that a quality
code includes phonetic and semantic information being optimally available to the working-memory processes. If semantic activation is insufficient, more effort is required for proposition assembly. If phonetic information is insufficient, the reference needed by memory may be
unavailable.

Actually, the general problem of code quality can be seen by considering *code asynchrony.* The assumption is that the longer a lexical process takes in its total execution, the greater the possibility that the activation of its component codes will be *out of phase.* In particular, phonetic
and semantic codes activated during access may be out of phase with
each other. To consider this possibility, recall the argument concerning

automatic activation of phonetic and semantic codes (Chapter 4). If access is rapid, all codes are rapidly available and the semantic code can immediately enter into proposition assembly. However, if semantic activation has peaked prior to activation of the phonetic code necessary to secure reference, then the overall quality of the word code is reduced. Or the opposite could happen—early phonetic activation and late semantic activation. In either case the longer total access time has made it somewhat more likely that a fragmented word code is available.

This asynchrony may be a way to understand two problems sometimes observed with low-ability readers. Immediately after reading a word in a text, the reader may be able to repeat the word but not "know its meaning"—even though he "knows" its meaning in some other situation. Or, immediately after reading a word in a text, he may have a vague memory for its gist that may result in recalling a semantically related word rather than the word itself. Or, on some occasions in oral reading, the "wrong" word is actually produced, reflecting semantic activation and phonetic-semantic asynchrony. Because of the asynchrony, the activation of the phonological form of the "wrong" word is momentarily higher than the actual printed word. This can occur occasionally in high-ability readers as well. Moreover, something very like this asynchrony is quite striking among "deep dyslexic" patients who, for example, misread the word *negative* as *minus* (Patterson & Marcel, 1977).

A related implication of the asynchrony hypothesis is that the context-appropriate meaning is less active. Recall that all meanings of a word may be activated (the lexical autonomy principle), before the required one is selected. However it is quite likely that normal texts do not produce so much semantic "priming." In such a case, all possible meanings should retain activation long enough to allow following context to select the correct one. Semantic asynchrony may put all meanings at a lower level of activation at exactly the moment that the context can make its best contribution.

Intrinsic Memory Hypotheses

The lexical access hypotheses attempt to explain how something goes wrong in working memory because of inefficient lexical processes. An alternative explanation is that lexical processes execute their work efficiently enough but that the efforts of working memory itself are at fault. For reasons lying beyond lexical access, working memory fails to hold onto words or to the propositions it is encoding. There are two general possibilities. One assumes a *functional* memory problem that is especially vulnerable to propositional encoding (ordered language coding generally). The other assumes a structural memory problem, a quite general capacity limitation.

Propositional encoding

The major alternative to the lexical access hypothesis within verbal efficiency theory is that some process intrinsically responsible for propositional encoding produces ability differences. The assumption is that lexical access processes have achieved a word code with relatively high efficiency. That is, several words have been efficiently accessed but the demands of propositional encoding are not met efficiently. For example, integration of propositions may be a problem because the reader makes unnecessary memory searches trying to make links that should be made automatically in short-term memory. Another possibility is that the reader has to make too many *active* inferences because schemata are not activated efficiently. Perfetti and Lesgold (1977) discuss the possibility that schemata can be optimally "bound" to individual words. For instance, in the *opened the window* example, the reader does access both *open* and *window*. However, this should result in a schema that binds *open* to *window*, namely the window opening schema. In this case, it binds *they* as well. It is efficient to have the schema variables (the agent and the recipient of *opening*) specified as part of the schema, so that they can be used in the next sentence. For example, if the next sentence says something about the window breaking, or Joe sticking his head out the window, etc., these facts are easily comprehended because the window-opening schema is available. On the other hand, if too much information is activated with the schema, inefficiency would result, with the addition of useless information to the memory representation.

Individually, these proposition-encoding possibilities seem a bit tentative, and in fact they are difficult to demonstrate clearly. However there is a common underlying process description that has plausibility. It is the *manipulation of language codes* in memory that is the major source of efficiency differences. Once they are activated, words and their components are vulnerable to memory loss. Two possible memory loss mechanisms were described by Perfetti and Lesgold (1977) as *hysteresis* and *specificity ordering*. Hysteresis is the possiblity that memory mechanisms are insufficiently labile (ready to change). The memory system is still working on one word—assembling a proposition for it—when the next word enters the system. It fails to react in synchrony, much like the semantic asynchrony possibility already discussed. The *specificity-ordering* possibility is that the quality of the code is low. Its specific semantic properties and its ordering position are vulnerable. These two possibilities describe things in complementary ways, one emphasizing the *system's* reactivity (hysteresis) and the other the *code's* quality.

Notice that these two possibilities are very close to describing a lexical access problem. They refer to things being out of phase in memory or to codes being of low quality. This reflects an important fact about processing descriptions of this kind. These are overlapping and interdependent processes. If a low criterion for lexical access is accepted in order to have rapid access, it is possible that some of these memory differences will be observed.

Short-term memory capacity
The simplest memory mechanism that will produce efficiency differences is the capacity of short-term memory. We have pointed out that reading ability differences can exist in the absence of demonstrable differences in a passive primary memory. However they will surely exist when such structural memory differences are present. This is a simpler explanation and needs no recourse to such things as asynchrony. It just stipulates a system that cannot hold enough information. In a sense, such problems are beyond the domain of verbal efficiency theory. A system-limited reading process, by definition, would be operating efficiently if its structural capacity is being used fully. Even so, there would be the question of whether we should be content with a mechanism solely dependent on abstract capacity. That too might deserve explanation in terms of function.

Linguistic Code Manipulation

It is clear that any of the above hypotheses is a reasonable elaboration of verbal efficiency. Of course, none of them is necessarily correct. They are simply logical possibilities consistent with the general assumptions of verbal efficiency.

Are they equally plausible? When the salient facts of reading ability are considered, there are grounds for distinguishing among these hypotheses. One salient fact is that lexical access and decoding differences are pervasive. Memory differences among readers are not generally found in isolation from lexical and decoding differences. This fact favors a lexically based explanation of efficiency.

However, a second salient fact of reading ability is that ability differences are often found for speech processing as well as print processing (for example, memory for spoken language and memory for printed text). This fact seems to favor an intrinsic memory elaboration of verbal efficiency. Alternatively, it requires a lexical explanation that is indifferent to input modality.

Thus, if there is to be a single mechanism, it will have to accommodate two facts that are, superficially at least, not compatible. However, if we assume a general linguistic coding process, things fall into place. The essential characteristics of this linguistic coding process are as follows: *In an ideal system, an inactive memory responds to a linguistic symbol, in any modality, by a rapid retrieval of codes that are part of that symbol's memory location.*

To the extent that these codes are retrieved rapidly and are high in quality, the system is efficient. To the extent retrieval is effortful and the retrieved codes are low in quality, the processing is inefficient. Since effort and speed have already been discussed, only a sense of code quality is needed.

Code quality. A linguistic code is high in quality to the extent that it contains both semantic information and phonetic information sufficient to uniquely recover its memory location.

The code does not have to be high in quality for a long period of time. It must be of high quality long enough for subsequent processes to perform their work. Thus a "name" without meaning and a meaning without a "name" are both low quality.

Notice also the assumption of an inactive memory. That means the memory system has not been primed. This retrieval from inactive memory is a critical feature of high levels of reading ability. Thus, *effortless retrieval* from inactive memory (effortless retrieval is "activation" rather than retrieval) and a *high-quality code* are the two essential features of the concept of efficient linguistic coding. The manipulation of such codes in memory—first their retrieval, then their manipulation in memory—are the essential local processes of reading.

Can this linguistic coding concept explain the two basic phenomena of pervasive ability differences in lexical access and modality-independent ability differences in memory (at least in many cases)? The second phenomenon is easily explained, because a system with low-quality codes or one that expends effort on code manipulation will show general memory problems, independent of input. The first phenomenon—pervasive ability differences in lexical access—requires further study. One possibility is that lexical access differences studied in print will also be found in speech recognition. Assuming any such differences are either negligible or smaller than differences in reading, this fact is explained as follows: Lexical access from print (or decoding from print) depends on linguistic code manipulation more than does access from speech. For the low-ability reader, lexical access from print, in effect, demands an early process of code translation. This process, at minimum, involves access to and manipulation of linguistic symbols at the letter and word level. The simplest explanation is that activation pathways between the letter level and the word level are not as easily activated (this is the model described in Chapter 2). However, a fuller account will have to consider activation of subword units, i.e., orthographic and/or phonetic pathways to the word. In other words, visual word recognition requires from the low-ability reader more of what he is not good at (retrieval and manipulation of linguistic codes) than does aural word recognition.

Part of this picture of verbal efficiency depends on phonetic codes as well as semantic codes being retrieved and manipulated. As described in Chapter 4, we assume that phonetic codes are part of what gets activated during lexical access. A system low in verbal efficiency, one that does not rapidly retrieve a high-quality code, may produce either a slower activating phonetic code or a lower quality one.

It is possible to link the concept of the efficient linguistic coding process to the various elaborations of verbal efficiency previously discussed. That is, an inefficient linguistic coding process may inhibit verbal efficiency in three ways: (1) lexical access interference, resulting from an inefficient code retrieval that places verbal memory at risk; (2) a low-quality code, when the inefficient linguistic coding operation either fails to activate all the needed semantic and phonetic codes or a high-quality code is at risk because of competing memory demands, (3) incomplete propositional encoding, when functional working memory is limited by inefficient code manipulation. Thus the three elaborations of verbal efficiency each predict processing "hitches" that might be observed under any particular situation. What seems to be the best description in any particular case will depend on how the reader is trading off the costs of his inefficiency.

Finally, verbal efficiency theory does not imply that the reader is totally at the mercy of low-level processes. It assumes that the component processes of reading are interactive and that there are procedures that can be applied to compensate for inefficiency of low-level processes. For example, if relevant schemata are made more available, they can help the reader integrate propositions, compensating for inefficient propositional encoding. Similarly, inefficient lexical access can be overcome, to some extent, by schema-based processes.

SUMMARY

Reading ability has many different components. Verbal efficiency theory attempts to give a comprehensive account of reading ability, defined by comprehension, by focusing on the resource cost of different components. It assumes that general constraints on memory and attention place a premium on efficient processing. It assumes that the reader can reduce, through learning and practice, the resource demands of some processes. It also assumes that allocation of resources contributes to efficiency.

The core proposal is that comprehension is limited by the efficient operation of local processes. Efficiency is defined as the quality of processing outcome in relation to the cost to processing resources. The various components of reading vary in their potential for efficiency. Schema activation and lexical access are potentially very low in resource cost, while proposition encoding is high.

These aspects of efficiency are demonstrated in the text work necessary for understanding a brief text segment. The various components—lexical access, propositional assembly, propositional integration, and text modeling—are overlapping (cascade) processes. Most of the text work goes to propositional encoding in the ideal case. Ability differences and text differences (readability) are described in similar terms, in reference to the text work components.

The more specific components of verbal efficiency depend on how observed differences in local processes are explained. One possibility is that lexical access inefficiency is responsible for observed differences in memory as well as lexical processes. The other is that intrinsic working-memory differences are responsible. The salient facts may be explained by a generalized *linguistic coding* process that affects the speed and quality of both the lexical access and the manipulation of codes in memory. An inefficient linguistic coding mechanism would produce access interference with the current contents of memory, lower quality word codes, and intrinsic linguistic memory problems.

NOTES

1. The definition of automaticity will turn out to be noncritical. The usual understanding of "automatic" implies a process requiring no resources. One test of whether a process is automatic is whether a second task can be performed at the same time without the performance decrement that would result if resources had to be allocated to the process thought to be automatic. It is difficult to say whether any process can really meet this test—certainly not processes as complex as those discussed here. Hence, the quotation marks around "automatic." An "automatic" process here is merely one that is less resource costly.

2. Kintsch and van Dijk (1978) assume that, when possible, a proposition is linked with propositions previously stored in a short-term memory buffer. This is done by argument overlap, i.e., shared reference between sentences, at no cost to resources. When a match is not found in the buffer,

a search of memory is initiated. The *integrative* process I refer to are those mainly based on matches found either by active memory search or by reinstatement. Thus, argument overlap is assumed to be nonintegrative except when anaphora is involved. The reason for not simply using Kintsch and van Dijk's (1978) terminology is to leave open the question that some integration work may be necessary even when argument overlap is found in the buffer. For example, there is something intuitively more integrative about a causal proposition that has argument overlap compared with two adjectives that are linked to the same noun. For example, a *warm* room and a *stuffy* room may require less integrative text work than *because it was hot* even when the linking arguments are in the short-term memory buffer in both cases.

3. Partial assembly of propositions, as well as complete assembly, depends on tentative groupings of encoded words. How a reader arrives at these groupings is the "parsing problem" of sentence comprehension. Solving the parsing problem is made difficult by the explosion of parsing possibilities that arise and the demands on resources that would be required to keep track of each one. Strategies that people use in parsing sentences to reduce the work have received quite a bit of attention (Frazier & Fodor, 1978; Frazier & Rayner, 1982; Kimball, 1973). For many texts, parsing is a very resource-costly process.

4. This way of representing several overlapping processes was first proposed by McClelland (1979) for word identification. This concept of "processes in cascade" is useful for describing any set of processes in which some processes are logically prior to others, but for which completion of a logically prior one is not necessary. It is essentially a form of a weakly interactive model. Process 2 needs *some* information from process 1, but it does not wait for all of it. Perfetti and Roth (1981) suggest that this cascade example can be applied to processes beyond word recognition, essentially the point being made here. It is interesting to consider the possibility that the cascade model, originally applied to word recognition, is actually better applied to text comprehension. In text comprehension it is unreasonable to assume several different levels of information that can activate each other without constraint. Thus it is perfectly reasonable to suggest that a fully interactive model, i.e., one with unrestricted downward activation, is appropriate for word recognition while a constrained model dependent on upward activation is required for text comprehension.

5. It is of practical interest that the lower ratio of propositions to words is approximately captured by lexical density, a measure of the percentage of a sentence's words that are "content" as opposed to "function" words. This of course reflects the fact that content words must each have at least one proposition for each occurrence, and function words often do not (Perfetti, 1969). Lexical density can be roughly used as an index of propositional density.

6. Propositional assembly is partly a function of the density of propositions and of new arguments. It is also a function of sentence parsing, probably the most important process ignored by text-processing models such as Kintsch and van Dijk (1978). However, there are obviously other potential factors: the number of arguments normally in a proposition, the differences among attributive (adjective) propositions and action and state

propositions, etc. For example, propositions may involve no extra work when one of the arguments is implied by the schema for the verb rather than explicitly stated: *Type a letter* (with a typewriter); *see a news report* (on television). The implications for ability are interesting. Is proposition assembly made more difficult by inserting words that can be easily inferred?

CHAPTER SEVEN

The Evidence of Reading Ability and Verbal Efficiency

This chapter examines some of the research on reading ability in the light of verbal efficiency theory. Because the body of research is large and sometimes inconclusive, this will not be a comprehensive survey. Instead it will selectively discuss evidence that seems to bear on the issues of verbal efficiency. These include prominently the following: (1) the relationships among general reading comprehension ability, local text processes, and lexical access and (2) the basic abilities underlying verbal efficiency. These issues are addressed by research that has compared children of different reading ability or otherwise directly investigated reading ability.

Most of the research discussed in this chapter was carried out at the University of Pittsburgh. Referring mainly to this research allows general conclusions to be drawn across different experiments that have used strictly comparable subject populations and identical definitions of reading ability. Such a procedure will allow for clarity and continuity of discussion without excessive detail and caveats. Thus there is some other important research that will not be discussed until later chapters.

THE PITTSBURGH RESEARCH

In Chapter 5, the general processing factors that contribute to reading ability were identified. The evidence there was largely based on the Pittsburgh research, studies carried out on children from grades two through six in schools of an urban Pittsburgh neighborhood. Demographically, the neighborhood is largely white and working class. The mean IQ of the low-ability readers considered in these studies is above 100 (no one below 95), and usually the IQ score of the low-ability reader is not much lower than that of the typical high-ability reader. They are mixed-sex groups with low-ability readers, on average, about 1.5 years below the grade level for their age on the Metropolitan Achievement Test (MAT) reading and comprehension test.

The evidence from this population described in Chapter 5 identified ability differences in memory and lexical access, including decoding. Here we will first describe evidence that links differences in these processes more directly to comprehension. Then we will describe evidence that tries to elaborate the nature of local processing differences.

Comprehension

High-ability readers, by definition, are better at global reading comprehension than are low-ability readers. For verbal efficiency theory, it is necessary to demonstrate that differences in comprehension occur in local text processing. Two previously unpublished studies will provide the needed demonstration.

Sentence verification

What is the simplest comprehension event? Aside from semantic encoding of individual words, it is the processing of a minimum sentence in which the reader compares the sentence input with his or her memory. *Sentence verification* is the name of this process: The subject decides whether a sentence is a true description of the world.

To make sure that *minimal* comprehension demands were present, we constructed very simple sentences requiring only semantic category information, e.g., *An apple is a fruit.* Verification here is essentially a question of comparing semantic information in memory with a semantic representation of the sentence. A processing model for such a sentence is as follows:

1. Encode proposition: *is (apple, fruit).*
2. Retrieve information from memory: *is (apple, fruit).*
3. Compare representations: match → yes, mismatch → no.

This model is of the class developed in sentence verification research by Clark and Chase (1972), Trabasso, Rollins, and Shaughnessy (1971), and Carpenter and Just (1975). Of course, there were also false sentences and, to make things complex once in a while, negative sentences. The resulting four sentence types are shown in Table 7-1.

The Experiment. There were two lists of sentences of the type shown in Table 7-1. In total there were 64 sentences representing 16 semantic cat-

TABLE 7-1 Verification Experiment: Sentence Examples and Mean Verification Times

Type	Example	Mean verification times (in sec)		
		Skilled	Less skilled	Difference
True affirmative	An apple is a fruit.	2.70	3.67	.97
False affirmative	An apple is a sport.	2.92	4.26	1.34
False negative	An apple is not a fruit.	3.20	4.90	1.70
True negative	An apple is not a sport.	3.35	4.83	1.48

egories, with each list containing 32 sentences and 8 categories. The 16 categories, with the positive instance used were as follows: *bird (sparrow), building (hotel), color (purple), sport (hockey), fruit (apple), metal (copper), tool (hammer), insect (cricket); animal (monkey), tree (maple), flower (tulip), furniture (table), month (August), vegetable (radish), city (Pittsburgh), clothing (sweater).* (The first 8 are from one list and the second 8 are from the other list.) False affirmatives, true negatives, and false negatives were constructed by pairing each category with two other instances and each instance with two other categories from the same list.

The subjects were 24 fourth-grade students. The average low-ability (less skilled) reader had a reading grade-equivalent of 2.6 and was at the 16th percentile of the MAT reading test (range 8–22). High-ability (skilled) readers were from the 76th percentile (range 60–92).

Results. The mean decision times for the four sentence types are shown in Table 7-1. One important result is that high-ability readers verified even the simple true affirmative more quickly than low-ability readers. When we consider that this is a simple five-word sentence containing only two content words, the difference of 970 milliseconds is rather substantial. What does this time include? It includes the time to access the words, assemble the proposition, retrieve information from memory, and compare this information with the proposition. But it does not include any of the difficult mental work of mismatch and negation that the other sentences require.

Table 7-2 illustrates the kind of model commonly assumed to describe the verification process. It assumes that the initial processes of (1) assembling a proposition from the sentence and (2) retrieving the information from memory are followed by (3) a process that compares the results of (1) with (2). Two processes may add more time: a mismatch in this comparison stage *(M)* and the encoding of a negative *(n)*. The encoding of a negative *always* leads to a mismatch when an affirmative memory representation is compared with a negative encoding. Thus *n* is really a combination of negation encoding and negation mismatch.[1]

TABLE 7-2 Verification Predictions According to Successive Comparisons Model

Type	Sentence encoding	Memory representation	Comparison processes	Predicted time
TA	+ (apple, fruit)	+ (apple, fruit)	match	t
FA	+ (apple, sport)	+ (apple, fruit)	mismatch	$t + M$
FN	NOT (apple, fruit)	+ (apple, fruit)	match negation	$t + n$
TN	NOT (apple, sport)	+ (apple, fruit)	mismatch negation	$t + M + n$

Note. The four sentence types are shown in order of increasing decision times from *TA* (true affirmative) to *TN* (true negative). The assumptions are consistent with a family of models first described by Trabasso, Rollins, and Shaughnessy (1971) and Clark and Chase (1972). It assumes that a sentence is verified by comparing its propositional representation with a semantic link (apple is a fruit) retrieved from memory. Processing time is added when there is a mismatch of the representations *(M)* and when a negative is encoded *(n)*. The presence of a negative always neans that a second comparison process occurs. It is assumed that the embedded proposition is compared first, then the negative element.

As Table 7-2 indicates, a false affirmative should take longer than a true affirmative by time M, the mismatch of a basic memory comparison. A false negative does not have this mismatch but does have a negative encoding and mismatch. And a true negative has both M and n, since M occurs as a mismatch between the basic propositions and it is followed by a negation mismatch. So, by this model of negation (called the "true" model by Clark & Chase, 1972), a true negative requires more time than any other type.

These modeling assumptions were applied to the data of Table 7-1 (actually to the mean data from individual subjects). For the high-ability readers, data of eight of the twelve were well predicted by this model. For the low-ability readers, the model was a good fit for only four of twelve subjects.[2] The conclusion from the modeling is that most high-ability readers conform to the verification model but that low-ability readers are more variable. It is the kind of variability that reflects inefficient process. There appear to be nonsystematic additional processes, such as rechecking, among the low-ability readers.

Actually, it is clear from Table 7-1 that low-ability readers were more affected by sentence complexity, whether it was negativity or mismatching. However the comparison between true and false affirmatives is the most enlightening. It requires no assumptions about how negatives are processed. The differences between the groups is quite dramatic. High-ability readers required only 220 additional milliseconds for falsification. Low-ability readers required an additional 590 milliseconds.

Interpretation. Why is there an extra penalty for the low-ability reader who must falsify a sentence? It is a demonstration of the *inactive memory* principle of verbal efficiency. In an affirmative sentence, semantic activation spreads from *apple* to *fruit*. *Fruit* is accessed more easily. An *apple* is a *sport* does not allow such activation. *Sport* has to be recognized on its own. Of course this also affects assembling the proposition: *Is (apple, fruit)* is easily assembled; *Is (apple, sport)* is not. Thus, the low-ability reader takes longer for a false affirmative because it does not match his memory representation.[3]

In summary, the verification experiment provides validation of a central assumption of verbal efficiency. Low-ability readers not only comprehend texts less well, they require more time to process the simplest sentences. The fact that they take even longer as the sentence adds some processing complexity is consistent with the assumption that manipulation of word codes in memory is the source of the difficulty, either at the level of simple lexical access or at the level of semantic encoding.

Left unanswered by the sentence verification study are two interesting questions. First, if low-ability readers require more processing when they have to read only a single sentence with two content words, what happens when sentence integration is involved? Also, the suggestion has been that simple lexical access might be responsible for the longer comprehension times of low-ability readers. But semantic word encoding or propositional encoding could account for the added time. Can evidence be brought to bear on these possibilities? The verbal arithmetic experiment was intended to help with both of these questions.

Verbal arithmetic
Verbal arithmetic refers to the solving of so-called word problems. A typical example of a word problem is this:

1. *Susan had seven apples.*
2. *Then Jim gave her two apples.*
3. *How many apples does Susan have now?*

The arithmetic in this problem is completely trivial: $7 + 2 = 9$. What makes it interesting for reading is that in order to solve it, the subject has to read, at least a little. There is another important feature of verbal arithmetic: It allows something to be said about what the reader's text model might be during processing. At least it allows the assumption that the reader will have constructed the correct model of the text only if he responds correctly concerning the number of apples or whatever objects the problem refers to.

A semantic representation system underlying word problems has been described by Heller and Greeno (1978). They distinguish among three basic semantic structures in word problems, which they refer to as *move*, *combine*, and *compare* structures. The *move* structure is the one we have studied. It basically describes a series of three states that change in the problem: (1) Some person *A* has some number of objects *X*. (2) Some person *B* causes a change in this number by an amount *Y*. (3) In the resulting new state, *A* now has *Z* objects. Depending on which of six problem types the subject reads, he can be asked to produce the result of the final state *(Z)*, the number of the initial state *(X)*, or the number of the change *(Y)*. Further, the change of state from *X* to *Z* can be either an increment (+) or a decrement (−). Examples of the six problem types used in the experiment are shown in Table 7-3.

Now suppose that children are given simple *move*-structure word problems to solve. We can certainly expect low-ability readers to take longer than high-ability readers. However, we can do something with the problems to make them lexically interesting. If we change certain words from one problem to the next, we are putting slightly greater demands on lexical access. We know that the repetition of words increases the ease with which they are processed, both within a word list and within a text (Dixon & Rothkopf, 1979). For the word problem experiment, we changed the two exchange agents, the exchange object and the exchange verb. For example, *Susan* and *Jim* could be changed to *Mary* and *Fred*, and *apples* could be changed to *marbles*. So in one condition, six consecutive problems were all about Susan giving Jim some apples or Jim giving Susan some apples. In this lexically *constant* condition, each problem required a different solution type, but the key words did not change. In the *lexically variable* condition, the same six problem types were encountered as in the constant condition, but the key words varied from one problem to the next *within* the set of six.

Experiment 1.[4] Subjects were presented with the entire problem visible at once on a computer terminal. They responded by pressing the single digit that was the correct answer. The difference between lexically *constant* and lexically *variable* was only in the distribution of the problems. The constant set was blocked, with each problem within a block of six repeating the same set of words. Subjects also performed a control task of number problems. The number problems require exactly the same computations as the word problems. For example $7 + 2 = ?$, $3 + ? = 8$, and $? - 2 = 4$ represent three of the six number problem types. There was one such problem corresponding to each word problem.

Subjects were 28 fourth-grade children, including a subgroup of 16 who provided a reading ability contrast, 8 high- and 8 low-ability readers. Low-ability readers averaged in the 26th percentile (range 2–48) and had an average grade equivalent of 3.0.

TABLE 7-3 Verbal Arithmetic Experiments: Problem Types and Solution Times

Problem type	Example	Mean solution times (in sec)			
		Unadjusted		Adjusted	
		High ability	Low ability	High ability	Low ability
$X+Y=?$	Susan had 7 apples. Then Jim gave her 2 apples. How many apples does Susan have now?	6.54	10.53	1.09	1.32
$X+?=Z$	Susan had 3 apples. Then Jim gave her some apples. Now Susan has 8 apples. How many apples did Jim give her?	9.35	13.60	.94	1.36
$?+Y=Z$	Susan had some apples. Then Jim gave her 4 apples. Now Susan has 5 apples. How many apples did Susan have in the beginning?	9.84	12.08	.89	1.10
$X-Y=?$	Susan had 8 apples. Then she gave 6 apples to Jim. How many apples does Susan have now?	7.59	11.71	.95	1.46
$X-?=Z$	Susan had 9 apples. Then she gave some apples to Jim. Now Susan has 5 apples. How many apples did she give to Jim?	8.74	12.61	.87	1.26
$?-X=Z$	Susan had some apples. Then she gave 2 apples to Jim. Now Susan has 4 apples. How many apples did she have in the beginning?	8.08	12.30	.73	1.12

Results. Table 7-3 shows the mean solution times, both for the problem as a whole and adjusted for the number of propositions. (Some problems had four sentences and some had three, resulting in a range of 8–11 propositions per problem.) Low-ability readers took longer to solve all types of problems. (Problem type differences are interesting but beyond our purpose.)

Table 7-4 shows the comparisons for high- and low-ability readers for the lexically variable and lexically constant problems, as well as for the number sentences. High-ability readers were faster at verbal problems than low-ability readers. They were faster also at number problems, but not significantly. (The two groups did not differ in accuracy for either word problems or number problems.)

The relationship among number computation speed, word problem speed, and reading ability can perhaps best be seen by using the data of all 28 subjects, who represent a wide range of reading skill. Numerical

TABLE 7-4 Mean Solution Times for Different Kinds of Problems

| | Solution time (in sec) | | |
| | | Verbal problems | |
Group	Number problems	Lexically variable	Lexically constant
Fourth-grade total sample	4.12	10.15 (1.05)[a]	9.42 (.97)
High ability	4.06	8.69 (.90)	8.02 (.83)
Low ability	4.70	12.81 (1.32)	11.46 (1.19)

[a]Numbers in parentheses represent the mean solution time per proposition.

computation speed correlated with word problem speed $r = .31$. A standard math achievement score that taps mathematical concept knowledge correlated with word problem speed $r = .27$. However, an independent measure of each child's vocalization latency—taken on different words—correlated with word problem speed $r = .72$. Thus speed of lexical access is the best predictor of speed of problem solution.

Finally, we can see the effect of lexical variability. Table 7-4 shows that for the fourth-grade total sample lexically variable problems required an average of .73 seconds longer to solve than lexically constant problems: Lexical variability affected only solution time, not accuracy. This is a small difference (about 8%) considering total problem solution time, but it does amount to about 80 milliseconds per proposition.

However, even this 80 milliseconds per proposition underestimates the impact of lexical variability. Its effect should be primarily in initial assembly of the propositions that contain the key words (the words that changed). Once these propositions are established, subsequent integration (based on argument overlap) should require no extra time. This means that the *first* proposition plus one other in the second sentence contain the effect of lexical variability. There are exactly two propositions affected regardless of problem type. Thus we can allocate the total solution time difference between lexically constant and lexically variable problems to just two propositions. This produces an estimate of 365 milliseconds per proposition, an increase of about 30%. This means that *lexical access*, prior to propositional encoding, can contribute substantially to reading times. Notice that that this is not a question of word difficulty based on length or frequency. Such variables would produce even longer differences.

The lexical variability effect is best explained this way: There is a cost to processing of lexical access that is difficult to avoid by use of semantic structures. The semantic structures in this case were uniform. It does not matter to semantic representation whether person A is named Susan or Mary or whether the exchange object X is apples or marbles. However, at least for fourth-grade children, these words must be accessed as names, not just used as semantic variables that fill in the problem schema.

As for skill differences in the effect of lexical variability, they were statistically unreliable for only eight subjects. However, it is clear from Table 7-4 that the means of low-ability subjects were affected more by lexical variability—130 milliseconds compared with 70 milliseconds for high-ability readers. Thus, it seems reasonable to conclude, first, that all readers are affected by lexical variability, because name code information is needed for problem representation. Second, since lexical access is more effortful for low-ability readers, they may be especially affected by lexical variability.

We now have evidence linking simple sentence comprehension to overall reading ability and linking simple text comprehension to lexical access as well as to overall ability. The second verbal arithmetic experiment was intended to shed some light on propositional integration in comprehension. If low-ability readers are weak at simple propositional assembly, as shown by the sentence verification experiment, this could be sufficient to account for any overall comprehension time differences such as we found in verbal arithmetic. On the other hand, it is possible that some local propositional processes are a source of extra difficulty. The second verbal arithmetic experiment explored this possibility by presenting word problems one line at a time. Since the text work varied from one line to the next, times to read individual sentences could be taken to reflect text work differences between high- and low-ability readers.

Experiment 2. The word problems of Experiment 1 were presented one line at a time to the same group of subjects. The subject pressed a bar to cause one line to be replaced by the next line of the problem. As shown in Table 7-3, four of the problem types had four lines and two had three lines. In addition, the problems necessarily vary in the kind of information provided by any given line. For example, in the first line, four of the six problems have a definite number quantifier while two of the problems have the indefinite quantifier *some*. And two of the problems allow computation during the second line, whereas four allow computation during the third line. However, every problem requires the *integration* of a state change in the second line with an initial state description of the first line. Thus the integration question can be asked at this point. Are there ability differences in second-line integration processes as well as in the encoding of the initial line?

Results. Low-ability readers required more reading time per sentence than high-ability readers. The main result of interest, however, is that there

TABLE 7-5 Reading Times for First and Second Sentences of Experiment
           2 Word Problems

Sentence	High skill	Low skill	(Means)
First sentence	2.36	2.84	(2.60)
Second sentence	2.64	3.52	(3.08)
(Means)	(2.50)	(3.18)	

was an interaction of sentence number and reading skill. As shown in Table 7-5, both groups of readers required more time for the second sentence. However, the extra time required by the low-ability readers was significantly longer than for high-ability readers, .68 seconds compared with .28 seconds.

Depending on the problem, some of these integrative sentences allowed computation of the answer and some did not. To make sure that the extra integration time required by the low-ability reader was not due to computation, we assessed each sentence of each problem, excluding the question, according to three general components: *integration (I)*, which applied to all sentences after the first; *computation (C)*, which applied to sentences that allowed numbers to be added or subtracted; and *basic encoding time* (T_0), which included lexical access, propositional encoding, response execution, etc., common to *all* sentences. When the reading times per sentence are partitioned into these components on an individual subject basis, skilled readers differ from less skilled readers in T_0 and I, but not in C. (For skilled readers, total reading time per sentence is $RT = 2.33\ T_0 + .26\ I + .12\ C$. For less skilled readers, $RT = 2.88\ T_0 + .56\ I + .12\ C$.)[5] Thus, it is linguistic sentence encoding and integration, not numerical computation, that is important.

Summary

Two experiments have been described to demonstrate a clear link between general reading ability and simple comprehension. The experiments clearly show that ability differences are present in the simplest sentence comprehension. The sentence verification results also show that minor additions to sentence complexity have large effects on low-ability readers. The verbal arithmetic experiments show that simple texts with known semantic structures produce reading time differences related to ability. The two groups of readers did not differ in the relevant knowledge but in their sentence-encoding processes, especially sentence integration. The studies also demonstrate the critical role of lexical access in a situation where top-down knowledge might be expected to dominate comprehension. Together with the studies showing immediate memory differences during text processing (Perfetti & Goldman, 1976; Goldman et al., 1980), they make the case that very local processing abilities contribute to overall reading comprehension ability.

Lexical Processes

In a sense, the results of simple comprehension experiments "reduce" global comprehension differences (based on answering questions in standardized tests) to "on line" processing factors. The clear suggestion is that what needs to be better understood is how propositions are encoded and integrated. As the discussion of verbal efficiency theory emphasized, there are several possibilities for a deeper understanding of such encoding differences. However, the facts concerning lexical processes are well established and a better understanding of lexical processes is clearly important.

The basic facts, to review them briefly, are that decoding and lexical access speed consistently distinguish high- and low-ability readers. Differences between groups are greater when linguistic codes must be processed, greater for longer words than shorter words, greater for low-frequency words than high-frequency words, and greater for pseudowords than real words. Such facts, coupled with the differences in simple comprehension, raise two complementary questions: What is the *effect* of lexical access differences and what is the *cause* of lexical access differences? The effect, according to verbal efficiency, is to increase the text work of propositional encoding. However this effect may act directly on propositional encoding or it may be mediated by inefficient semantic encoding. The causes of lexical access differences are equally complex. These questions are to be taken up in the next sections.

Semantic Encoding and Lexical Access

Semantic encoding

The question for semantic encoding seems to be this: Are low-ability readers less efficient in encoding the semantic properties of a word *or* are semantic encoding differences accounted for by simpler lexical access processes? If semantic features of a word are automatically activated by access, then the latter choice should be correct. If low-ability readers somehow fail to activate semantic codes and name codes simultaneously—the asynchrony hypothesis (Chapter 6)—then the first choice may be correct. It will turn out that this question is difficult to answer, partly because experiments have not been very sophisticated but partly because the question may be inappropriate. However, the attempts to answer it have involved tasks that try to separate decoding or lexical access from semantic decision.

One such experiment, described in Perfetti and Lesgold (1979), required fourth-grade subjects to perform matching tasks and categorization tasks. The matching task presented a word orally, followed by a printed word. The category task presented a semantic category orally, followed by a printed word. The subject had to respond "yes" or "no" in each case. The results showed a larger ability difference for category judgments than for word judgments. The same effect was found when the stimuli were pictures instead of words. The conclusion seems to be this: Very simple semantic decisions require additional time for low-ability readers, and this time difference cannot be accounted for by lexical access from print. A possible explanation is that for the low-ability reader, the semantic code is less strongly bound to the name code, as suggested by the asynchrony hypothesis.

Search experiments

Another way to separate semantic-level word encoding from name-level encoding is to use a visual search task. In such an experiment (described also in Perfetti, 1983) subjects were given a prespecified target followed by a visual display. In one condition the target was a specific *word*, and in another condition the target was a *semantic category*. The subject responded "yes" if the display contained the target and "no" if it did not. Since the display size ranged from one to seven words, estimates of processing *rate* could be obtained from the linear function relating search time to display size.

Examples of the materials for word targets and semantic category targets are given below for display size 5 (i.e., five words displayed).

> Word target present:
>   Truck: treat   trade   tune   truck   task
>
> Word target absent:
>   Truck: trip   touch   team   trash   tools
>
> Semantic category target present
>   Animals: fun   stuff   cat   past   low
>
> Semantic category target absent
>   Animals: call   peace   girl   good   view

Notice that the word-level task is not easy to perform visually. Because the nontarget words are similar to the target word in length and letter composition, readers may be expected to use lexical access processes that reach the name level. This should happen at least for readers who are efficient at this level of access.

Subjects in the experiment were 24 third-grade children. High-ability readers scored at the 77th percentile of the MAT, and low-ability readers scored at the 19th percentile, on average. Subjects participated in word-search, category-search, and pseudoword-search tasks in separate sessions. The target was blocked for eight successive trials for 16 word and category targets.

Results. High-ability readers were faster at both tasks than low-ability readers. For word targets these differences were present when the display size was one, and the differences increased with increasing set size. That is, there were both *intercept* and *slope* differences for word search. For category search, differences did not increase with increasing display size. Figure 7-1 shows the results for two subgroups, matched with respect to IQ, for positive (target-present) and negative (target-absent) trials.[6]

Word Search. For positive trials, high- and low-ability readers were not different in intercepts but were different in slopes, which were about 100 milliseconds faster for high-ability readers. For negative trials, the situation is similar, with a 180-millisecond slope difference. However, there is a slight but interesting difference between positive and negative trials. For negative trials, differences between high- and low-ability readers were clearly present for set size 1, as indicated by the means plotted in Figure 7-1 (right panel). This difference is about 200 milliseconds, quite comparable to the magnitude of word identification differences between high- and low-ability readers (Perfetti & Hogaboam, 1975). This difference between a *match* (positive trials) and a *mismatch* (negative trials) is quite interesting for our understanding of ability factors. It is reminiscent of the extra time required by low-ability readers for falsifying simple categorical sentences (see page 125). It is also reminiscent of the extra time in naming that a low-ability reader requires when his or her expectations for a particular word are not met (Perfetti & Roth, 1981). The principle we see operating in all these cases is the *inactive memory* principle. Low-ability readers are not so slow when the prior activation has occurred (when matching is possible). They are slower when prior activation cannot help (mismatches). Moreover, the fact that there were slope differences in word searches is interpreted as follows. High-ability readers have a faster *rate* of name-level word processing, not just a faster general reaction to a word display. This rate may be described as the time to *access* a word in the display and *compare* it with a word target in memory. These are two linguistic code manipulations that verbal efficiency theory claims are inefficient in low-ability readers.

Semantic Category Search. The semantic category data for these subjects are also in Figure 7-1. High-ability readers were faster than low-ability

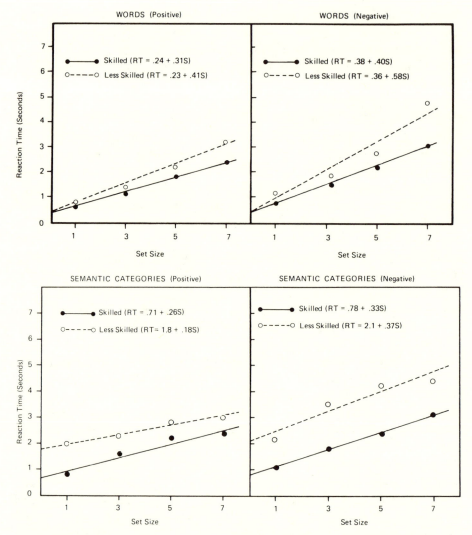

Figure 7-1. Search times for words (top panels) and semantic categories (bottom panels). Left panels show target-present conditions, and right panels show target-absent conditions. Best-fitting straight lines are shown for high- and low-ability third-grade subjects.

readers at all set sizes. Moreover, high-ability readers showed a good linear fit with low variances, whereas low-ability readers showed some nonlinearity and high variance. In order to avoid the problems of nonlinearity, we can make a comparison of semantic decision and word decision just for set size 1.

An index of semantic encoding beyond name-level encoding can be obtained by subtracting word decision time (WD) from semantic-category decision time (SD). This was done for each subject for set size 1. The resulting difference, SD − WD, can be taken as an index of semantic encoding controlled for word encoding. The mean for SD − WD for low-ability readers was 1,050 milliseconds, and the mean SD − WD for high-ability readers was 209 milliseconds. When we take this SD − WD difference over larger set sizes, the difference between high- and low-ability readers increases slightly from set size 1 (841 milliseconds) to set size 3 (950 milliseconds) and set size 5 (1,000 milliseconds), and then disappears altogether for set size 7. For the largest set size, both groups had essentially zero SD − WD differences.

The conclusion from such data is that some semantic encoding differences go beyond word-level differences. However, they are not rate-constant effects. They are large immediately when the subject has only a single word to encode. It is possible that most of the low-ability reader's work in this semantic task is the initiation of a search process for relevant semantic attributes. Once some semantic search process is initiated, differences in rate of semantic comparisons are smaller. Again in the low-ability reader we may be observing a linguistic memory system of high *inertia*. A single semantic encoding may overcome this inertia.[7]

Decoding and Lexical Access

The other side of the lexical access question is what produces the ability differences in access, regardless of their semantic consequences. Evidence seems to suggest that sublexical processes may be responsible.

Pseudoword and bigram search
The search experiments described in the last section were used to examine sublexical processes. In one task, subjects search for pseudowords such as the following, which illustrate the target-present conditions:

(short target) NUR: NAR   NUD   NIF   NUR   NEP
(long target) ZOBIT: ZEBUV   ZEDIK   ZOFEL   ZEMIR   ZOBIT

The target-absent trials were very similar. That is, the nontarget words were equal to the target in length and shared letters with it. Subjects searched 64 displays blocked such that eight consecutive trials tested the same target.

The results were that pseudoword search was faster for high-ability readers than low-ability readers. Ability interacted with length of the pseudoword and with size of the display. Low-ability readers were affected more by length of the pseudoword and by set size than were high-ability readers. Mean reaction times are shown in Figure 7-2 (top panel) combined over number of syllables. There are significant differences in slope as well as intercept. However, it is clear from Figure 7-2 that there are practically no differences between high- and low-ability readers for set size 1, especially for target-absent trials. This is very likely because of the usefulness of visual analysis at set size 1. The subject gets a good idea of what the target looks like and can reject a single item without decoding. As the display size increases, the value of decoding over visual matching increases and striking search differences emerge. The negative trials—where slopes are always the most interpretable since the subject must inspect each item before responding—suggest a *rate* difference of about 240 milliseconds attributable to ability. That is, it costs low-ability readers an additional 690 milliseconds for each additional pseudoword, compared with 450 milliseconds per pseudoword for high-ability readers.

Comparison of Figures 7-1 and 7-2 demonstrates the extra processing difficulty that low-ability readers experience with pseudowords. Using the slopes of the negative trials as rate indexes, we see that low-ability readers had a processing rate of 580 milliseconds per word and 690 milliseconds per pseudoword. High-ability readers had corresponding rates of 400 milliseconds per word and 450 milliseconds per pseudoword. This suggests that the extra cost of processing a pseudoword is greater for low-ability readers than high-ability readers, which agrees with name production data (Hogaboam & Perfetti, 1978) and implies a genuine rate difference as well.[8]

Finally, there was also a search task involving bigrams, the data for which are also shown in Figure 7-2. These data were collected on the same subject sample at a somewhat later time. Trial examples are as follows:

(target present) WP: MQ   WT   TL   WP   XP
(target absent) WP: MT   WL   SX   KD   MP

As the example indicates, the targets and displays were exclusively *consonants*. The principle was that the reader should search through simple displays of letters that were not codable as syllables. Some visual analysis was required by having some nontargets contain either the first or the second letter of the target. Clearly, however, the visual discrimination was

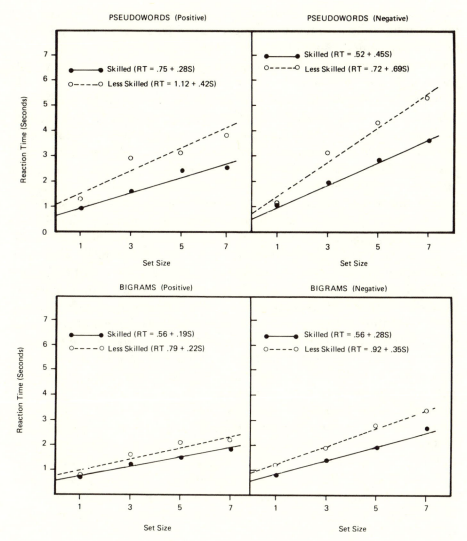

Figure 7-2. Search times for pseudowords (top panels) and bigrams (bottom panels). Left panels show target-present conditions, and right panels show target-absent conditions.

easy, compared with pseudoword search. Furthermore, the subject had blocks of 32 trials with the same target, making the task an easy one with no memory component and visually guided search.

On target-present trials there were only small differences between ability groups, and none at all for set size 1. (This latter result corresponds to the word data.) For target-absent data there were small but significant differences in intercept and slope. The slope estimates show that the rate per bigram was 280 milliseconds for high-ability readers and 350 milliseconds per bigram for low-ability readers. It is doubtful that this 70-millisecond difference in the rate of bigram processing can explain the much larger rate difference (240 milliseconds) for pseudowords, but a comparison of the two is difficult. In one attempt, a mean difference score between pseudoword times and bigram times was computed for each set size for each subject. Of course, pseudowords should take longer, merely on the basis of number of letters. However, these difference scores were much larger for low-ability readers than high-ability readers and they were especially large for larger display sizes. Moreover—and this is perhaps the most straightforward comparison—bigram search time could be compared with three-letter pseudoword search time. For low-ability readers the slope for three-letter syllables was 630 milliseconds, while for high-ability readers it was 409 milliseconds. Since this 200-millisecond difference is quite a bit larger than the 70-millisecond difference for bigrams, it suggests that ability differences do not depend on single-letter encoding, but on multiletter encoding.

Thus the conclusion from these studies is that there are genuine processing rate differences between high- and low-ability readers for words and for letter strings. The rate differences are especially large for pseudowords. It is possible that these rate differences cannot be reduced to simple letter encoding, although small bigram rate differences do exist. The question of the exact source of processing rate differences for letter strings remains open.

Orthographic structure
One possible source of the general word-processing advantage of high-ability readers is in the use of orthographic structure. By orthographic structure, we mean the patterns of grapheme sequences permitted in the language (see Chapter 5). If high-ability readers had greater knowledge of the orthographic structure of English or if they were better able to use it, this skill could help explain their advantage in word processing. And, equally important, it would help explain why their advantage is even greater for pseudowords than for real words. Pseudowords are essentially orthographically and phonetically regular letter strings.

To examine the role of orthographic structure, we had subjects perform forward and backward letter search through nonword strings. Nonword strings allow the effects of orthographic structure to be separated from other lexical processes. These search tasks have been used in other research on orthographic structure (Massaro, Venezky, & Taylor, 1979) and in other studies of reading ability differences (Katz, 1977; Mason, 1975; Massaro & Taylor, 1980).

In forward search, the subject is first presented with a target letter and then searches a visual display for the target. In backward search, the display is presented first for a brief, specified period and then it goes off. The target letter follows after a brief period. Thus forward search is visual search, and backward search is memory search.

In one experiment, 32 third-grade subjects performed both forward and backward search—16 high-ability subjects and 16 low-ability subjects.[9] Search strings were six letters long, controlled for their letter position frequency, i.e., how often a given letter occurs in a given position in English words. Strings high in structure were more pronounceable and more orthographically regular than were strings low in structure. For example:

High structure: NETOPE
Low structure: PBTCLA

In the forward search condition, a target letter preceded the display string, which remained on until the subject's response. In the backward search condition the display string was on for one second followed by a one-second blank interval and then the target letter. The main results were as follows.

Forward Search. Forward search performance was unaffected by structure and high- and low-ability subjects were not significantly different. An analysis of visual similarity between the target letter and the search string revealed that the main factor affecting performance was visual similarity. For example, in a target-absent trial, displays that included at least one letter with some visual feature overlap with the target, search was slowed by about 300 milliseconds. There was a comparable effect for a target-present trial. Thus forward visual search was mainly a visual process. Performance was affected by the visual feature similarity between targets and displays and not by orthographic structure. Importantly, high- and low-ability readers did not differ in the effect of this visual variable.[10]

Backward Search. The backward search task produced high error rates, and these rather than latencies were the most meaningful. The main results were that performance was affected by orthographic structure and that high-ability readers performed better than low-ability readers. Most interesting was that the effect of orthographic structure was greater for high-ability readers. High-ability readers had an error rate of 21.3% for high-structure strings and 31.5% for low-structure strings. Low-ability readers had a 40% error rate for both types. Thus the conclusion is that high-ability readers can take advantage of structure better than low-ability readers, at least when conditions put a premium on rapid encoding.

These results raise the following question: Are high-ability readers at an advantage in using structure in such a situation because of encoding or memory? To put it another way, will low-ability readers be able to use structure when they are given more time to encode the letter string or will the requirement to remember the string still cause a problem? One interpretation of the orthographic structure effect is that it results from the reader encoding the letter string phonetically for memory storage. The target letter is then compared with this phonetic code. Thus low-ability readers can be either inefficient in linguistic encoding or unable to retain the code in memory. These possibilities bear on the code-quality interpretation of verbal efficiency. That is, regardless of encoding, a low-quality code may often result for a low-ability reader.

A second experiment was carried out on subjects very comparable to those in the first study. This time there were two groups of low-ability readers, one of which was given 333 milliseconds to encode a six-letter string, while the other was given 1,500 milliseconds. A group of high-ability readers was given 333 milliseconds to encode the six-letter string. In addition, for all subjects there were two different letter probe delays. In the short delay, a blank interval of 500 milliseconds intervened between offset of the letter string and onset of the letter probe. In the long delay this interval was four seconds. Thus at the short interval, *linguistic* memory was trivial while *perceptual* memory was important.[11] At the longer interval, perceptual memory was less useful and linguistic memory more important.

The important results were these: (1) Low-ability readers given 1,500 milliseconds for encoding performed as accurately as high-ability subjects given 333 milliseconds for encoding. Low-ability subjects given 333 milliseconds were less accurate. (2) Even with the longer encoding time, low-ability readers were slower at responding when there was a four-second delay interval. Thus, reducing the demands for encoding efficiency did not completely eliminate memory inefficiency. (3) Low-ability readers with only 333 milliseconds to encode performed very inaccurately at the four-second interval. This is clear evidence for an encoding quality–memory trade-off. High demands on efficiency forced a low-quality code that was available after half a second but not after four seconds. (4) High-ability readers were unaffected, overall, by the memory interval. (5) All groups of subjects showed an advantage of orthographic structure under some conditions. For low-ability readers given only 333 milliseconds, structure was effective only after 500 milliseconds. For low-ability readers given 1,500 milliseconds, structure was effective only after four seconds. The explanation follows from the effects of encoding time and memory interval on code quality. With plenty of time to encode and a 500-millisecond memory interval, a high-quality perceptual code is sufficient. At longer intervals, a linguistic code is needed. With very little time to encode, the low-ability subject forms a code with some linguistic information but it is useful only for a brief time.

Overall, our conclusion is not that low-ability readers do not have knowledge of orthographic structure. They can use it in a letter search under some conditions. When either encoding demands or memory demands are minimal, low-ability readers show an advantage of structure. When both encoding and memory demands are high, their ability to use structure and their general performance levels both decrease.

Summary

The research on lexical processes, including semantic coding, supports this general picture: Compared with high-ability readers, low-ability readers are slower (and less accurate) at word access. One immediate consequence of their less efficient access is that semantic information stored with a word takes longer to get activated. There is some evidence from search tasks that semantic encoding requires more processing time for a low-ability reader and that ability differences in semantic encoding may not be completely accounted for by lexical access differences. The underlying components of differences in lexical access seem to include sublexical processes. Search data for nonword strings implicate processes that operate on orthographic structures, i.e., predictable letter sequences. However, at least some small processing differences occur at levels of letter encoding that do not involve structure. The use of structure, which may be mainly to allow linguistic coding (i.e., decoding) seems to be available to low-ability readers when processing demands are not high.

Lexical Processes in Context

One final question concerns the ability of readers to use context in lexical access. As discussed in Chapter 2, processes of lexical activation are influenced in some way by the semantic context. Although the exact mechanism by which context has its influence, and even the extent of its influence, are open to question, we make the following general assumptions. The activation level of a word's memory location can be increased by semantic processes initiated before the lexical access of the word through its visual features and letters has been completed. (This avoids misleading questions of whether these effects are "preaccess" or "postaccess.") We do not need a very powerful interactive model to account for such effects. One that simply allows semantic inputs to raise the activation level of a word (Morton, 1969) is sufficient. The processes of influence are best thought of as automatic and unconscious.[12]

Our experiments have not attempted to separate the automatic contribution of context from the attentional. They have assumed that in a task in which a subject has to read a word aloud in context, the context will exert a strong influence on the time taken to read the word. The time interval between offset of the context and onset of the word to be read was somewhat variable. In the one case in which this interval was under tight experimental control, it was about 500 milliseconds. This was a listening experiment in which the offset of a spoken text triggered the visual appearance of a word to be named (Perfetti, Goldman, & Hogaboam, 1979, Experiment 1). In other experiments, context was read by the subject, and an experimenter exposed the word to be named. This provided a more variable interval, but typically less than one second. The point of such detail is to make clear that we cannot say that the context effects we will describe are only automatic ones. Intervals of more than half a second may be enough time for conscious prediction processes to be effective and for unconscious activation to have weakened (Neely, 1977). However, it is possible that activation processes are much stronger in discourse and prediction processes much slower acting.[13]

In all experiments, the comparison is between the vocalization latency of a word in a discourse context and the latency of the word in isolation. The results of several experiments (Perfetti & Hogaboam, 1975; Perfetti, Finger, & Hogaboam, 1978; Hogaboam & Perfetti, 1978; Perfetti & Roth, 1981) can be summarized as follows: Both low-ability readers and high-ability readers name words faster in context than in isolation. This is true whether the context is read or heard by the subject and whether the context is a long story or merely pairs of sentences.

The word-identification latencies of low-ability readers were consistently more affected by context than were those of high-ability readers. For example, across the three experiments of Perfetti et al. (1978), the size of the context effect was 272, 200, and 314 milliseconds for low-ability readers. For high-ability readers, the corresponding context effects were 63, 41, and 92 milliseconds. Clearly, low-ability readers were helped more by context than were high-ability readers. There was also evidence that the effect of lexical variables—word frequency and word length—had less of an effect in context for both high- and low-ability readers. However, low-ability readers were more affected by frequency and word length than were high-ability readers, even in context.

Prediction

In one of the experiments of Perfetti, Finger, and Hogaboam (1978), subjects tried to predict the next word; immediately following their prediction, they had to identify the word. Two results are of interest in this task—the prediction accuracy and the effect of prediction on identification. In predicting the words, high-ability readers were more accurate at predicting the exact word, 32% compared with 22% for less skilled readers. Low-ability readers were also more likely to predict a word that did not satisfy the constraints of the sentence. The results of the identification were informative. Given that the word that was to appear was correctly predicted by the subject, subjects were faster at identifying it. However, a comparison between identification latencies to correctly predicted words and to words that were appropriate for the context but not predicted by the subject showed a difference between high- and low-ability readers. High-ability readers showed a gain of 50 milliseconds due to exact prediction, and low-ability readers showed a gain of 180 milliseconds. Clearly this context effect is one of conscious prediction, not automatic activation. This can be described either as a facilitation effect when the predicted word does appear, or as an inhibition effect when the predicted word does not appear. Either way, the effect of a successful prediction is larger for the low-ability reader.

This same kind of effect is seen in one of the experiments reported in Perfetti and Roth (1981). Subjects read short texts that ended with a blank denoting a word to be identified. These words varied in their predictability within these texts. One third of the words were predictable (80%), one third were not predictable (3%), and one third were anomalous (0%), as defined by the percentage of correct predictions of an independent group of subjects. For example, the word *cabbage* ended each of three texts. In the predictable context, it was preceded by cooking, as in ". . . *corned beef and* _____." In the unpredictable, it was preceded by planting a garden of ". . . *carrots, beans, and* _____." In the anomalous context, it occurred in a text about a woman checking her purse before leaving on a trip and finding "*a passport, a plane ticket, and* _____."

Third- and fourth-grade high- and low-ability readers were subjects. The data for the fourth grade are shown in Figure 7-3. While both ability groups were faster in predictable contexts, only low-ability readers were slower in anomalous contexts. Third-grade subjects showed a different pattern, one in which both ability groups were slowed down by anomalous contexts. (West & Stanovich, 1978, report similar differences between adults and children, with inhibition effects only for children.) The interpretation is that at least older high-ability readers have fast-executing lexical processes without context. The slower executing lexical processes of the low-ability readers make them more dependent on context. They show more facilitation when the word is predictable and more inhibition when

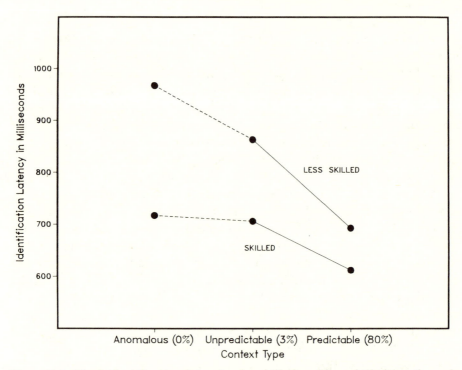

Figure 7-3. Word-identification latencies for skilled and less skilled fourth graders for three types of context. In parentheses is the average predictability in percent of the word to be identified. Less skilled readers were more affected by anomalous contexts. Data originally reported in Perfetti and Roth (1981).

it is not.[14] This general relationship between context facilitation and reading ability has been found in several experiments and can be summarized by Figure 7-4. The data in Figure 7-4 are combined over several conditions of visual degrading, which is why the mean scores are a bit high. (Without degrading, it is difficult to see much effect of context on high-ability readers.) The main thing to see is the greater effect of context on low-ability readers.

Context and basic word identification

The explanation for these kinds of context effects assumes that we may think of words and individuals in the same way as we described *texts* and individuals in Chapter 5. Each individual has a basic word-identification rate that varies over words. Each word has a basic identification rate (or activation rate) that varies over individuals. The individual rate is determined by individual skill, and the word rate is determined by word frequency, length, etc. A fast-rate reader identifies most words more quickly than a slow-rate reader. And a "fast-rate" word is identified more

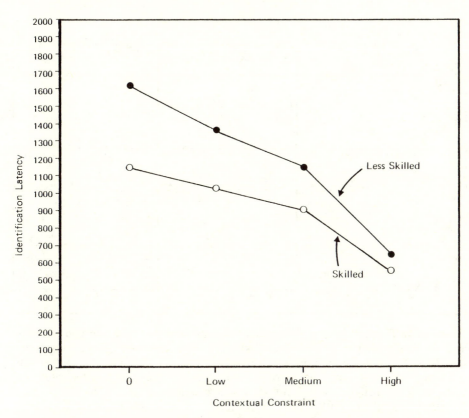

Figure 7-4. Less skilled readers' word-identification times are more affected by context than are those of skilled readers. Data are for fourth-grade subjects, averaged over conditions of stimulus degrading (Perfetti & Roth, 1981).

quickly than a "slow-rate" word for most readers. We further assume that the effect of context is the same for the basic word rate and for the basic individual rate: It adds information to the identification process, causing the word-activation process to execute more rapidly. These assumptions are illustrated in Figure 7-5. The top three panels show the word-activation functions for a fast-rate reader (or fast-rate word). The bottom three panels show the word-activation functions for a slow-rate reader (or slow-rate word). (The activation levels are expressed as percentages of the amount of activation needed for identification.) The effects of three different contexts are shown from left to right. The left panel is the identification function without context. The center panel shows the facilitation effect of a predictive context. The right panel shows the inhibiting effect of a misleading or anomalous context. In each case, the effect of context is t, the change in time units resulting from setting k, the context parameter, at a different value. As can be seen by comparing the panels,

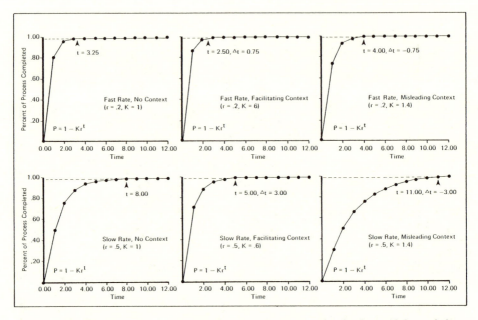

Figure 7-5. Theoretical word-identification functions for high- and low-ability readers. Top three panels show a high-ability reader, assumed to have a fast basic word-identification rate (r). Bottom three panels show a low-ability reader with a lower basic rate. From left to right, three types of context are illustrated: no context, facilitating context, and misleading context. The word-identification functions are the percentage of an activation process completed in arbitrary time units t. The value of t is the number of units prior to reaching asymptote and increases or decreases according to the helpfulness of the context. Low-ability readers are more affected by context than are high-ability readers.

a slow-rate reader is more affected by context than a fast-rate reader. Equivalently, a slow-rate word is more affected by context than a fast-rate word. These results were in fact found in the naming time data of Perfetti and Roth (1981) and in the data of West and Stanovich (1978). In the Perfetti and Roth studies, low-ability readers showed more facilitation for words named in context. And younger and less skilled readers showed more inhibition in anomalous contexts. Furthermore there were interactions of degrading with context. As degrading increased, so did the effects of context. This result was also found by Stanovich (1981). The general pattern from Stanovich's studies (1981; West & Stanovich, 1978) is that for adults, readers with fast identification rates, inhibition effects are found only for degraded words (words with slow identification rates). Inhibition effects for normal words are found in children rather than adults (Stanovich, 1981) and are larger for low-ability readers than high-ability readers (Perfetti & Roth, 1981). These facts are consistent with activation assumptions illustrated in Figure 7-5.

The assumption that words and individuals can be considered as "interchangeable" in their context effects also seems to be correct. For example, Perfetti and Roth (1981) manipulated the "activation rate" of a word by using different degrees of visual degrading. With degraded words, identification times of course are slower. This allows a comparison between high-ability readers and low-ability readers of the same "identification rates." That is, a high-ability reader can be found who, at 21% degrading, identifies a word at 1,200 milliseconds in isolation. And a low-ability reader can be found who has this same 1,200-millisecond rate but for a normal undegraded word. Now according to the assumption that context effects are rate constants, the effects of context on these high- and low-ability readers should be identical. They were. In a moderately constraining context, both high- and low-ability readers experienced about a 200-millisecond gain. In fact, there is a linear function relating the *facilitation* due to context to isolated word-identification time, independent of reading ability and word degrading. In other words, it does not matter whether a word's isolated identification time is measured from a high-ability or low-ability reader or from a degraded or nomal word. The context effect simply depends on the basic word-identification time. Context effects are greater for slower word rates, whether these rates are for words or for individuals.

Context in normal reading
Finally, we reconsider briefly the question of whether context effects actually occur in reading. The controversy over this issue was discussed in Chapter 2. The issue turns on whether normal reading occurs too rapidly to allow contextual processes to operate (Mitchell & Green, 1978). The implications of this possibility for reading ability are interesting. For one thing, it would imply that the kind of research referred to in this section does not tell us much about ability differences in use of context in actual reading. However, it may leave intact our major conclusion: that word identification of high-ability readers is less affected by context than word identification of low-ability readers. Even if the argument that reading is too fast for the use of context can be sustained, it certainly would not apply to children low in reading ability. Their slow rate in silent (and oral) reading is one of their salient characteristics. However, as we noted previously (Chapter 2), it is not clear that such an argument *can* be sustained. There are effects of context in experimental situations in which the time lag between the contextual event and lexical access is zero (Stanovich, 1981). Even more significant, perhaps, is the fact that context effects do not depend on the reader's ability to identify the contextual event (De Groot, 1983; Fowler et al., 1981; Marcel, 1983). For example, a word that is presented too briefly to be identified can nonetheless facilitate the

recognition of a related word that is presented immediately afterward (Marcel, 1983). The Posner and Snyder (1975) theory gives us a way to understand rapid and unconscious effects of context as well as slower and attentional ones. This theory has a fair amount of support for reading in context (Stanovich, 1981), and it is possible that the kinds of ability differences we have identified in the use of context apply to both automatic and conscious processes. The fact is, neither of these is well understood in the case of normal reading. For now, the application of studies of word naming to normal reading is reasonable, although it will be important to learn more from other reading situations.

A final observation is that if *active* prediction is significant in normal reading, this is something high-ability readers will do much better than low-ability readers. When subjects are asked to predict the next word in a text, high-ability readers are better than low-ability readers. This is of course to be expected. It is not uncommon to *define* comprehension as the success in predicting words deleted from a text (the "cloze" procedure). Although low-ability readers are just as good at prediction when the contexts are highly constraining, they are less accurate in more loosely constrained contexts (Perfetti and Roth, 1981). This fact, at first glance, may seem incongruent with the low-ability reader's greater dependence on context. However, it is actually quite compatible. The lower predictive abilities of low-ability readers are quite sufficient to be helpful in their reading, so they use them. Their basic word-identification processes are slow acting. Before these processes have enough time to execute, active predictive efforts are having an effect. Thus it is completely consistent that an individual whose contextual abilities are limited should use them heavily.

Conclusions from the Pittsburgh Research

In this chapter, we have drawn primarily on research from a particular population with uniform sampling characteristics from one study to another. This applies as well to much of the research described in Chapter 5. The research gives general support to the verbal efficiency theory and suggests some underlying processes that may contribute to efficiency. Here are the main conclusions.

Comprehension. Ability differences tapped by global reading comprehension tests are observable in "on line" comprehension. The time to evaluate a simple two-argument proposition (a sentence with two nouns linked by *to be*) distinguishes high- from low-ability readers. In addition, low-ability readers are slowed down dramatically by the addition of one extra process, e.g., the need to process negation or falsification. When three-sentence texts with word problems are read, low-ability readers are slower on a per sentence basis, but are especially slowed down by sentences that demand *integration*. The conclusion is that elementary propositional processes that must be repeated many times in text comprehension are less efficient in low-ability readers.

Memory. Our conclusion (from Chapter 5) is that the inefficiency of low-ability readers in propositional encoding is reflected in reduced ability to remember words just read. The forgetting increases when the words to be recalled are from a prior clause or sentence. This suggests a link between integration of propositions and memory loss for words needed for this integration. However, this word memory factor does not explain propositional comprehension directly, but only through a simpler linguistic coding mechanism. That is, the manipulation of words in memory is the key. (Considered simply as "word memory" it cannot explain why simple verification involving two words should be difficult.)

Lexical Access and Semantic Encoding. Low-ability readers require more time to access a word and to activate its name code. These are the pervasive facts from many studies. One consequence of this slower activation is clearly a more slowly activated semantic code. Some evidence was presented that suggests that semantic encoding may constitute an additional inefficient process of the low-ability reader, rather than being wholly accounted for by the difficulty of lexical access. In either case the semantic coding difficulty provides a needed link to propositional encoding.

Lexical Access and Decoding. It is clear that the processes of lexical access are important to describing the inefficiency of the low-ability reader. For this population, lexical processes beyond general name retrieval are implicated. Evidence suggests that the relevant processes include sublexical processes, not just "holistic" word processes. Processing rates for pseudowords show especially large ability differences. One important sublexical process may depend on use of orthographic structure. Evidence suggests that high-ability readers can take advantage of this structure under more circumstances than can low-ability readers. However, low-ability readers do have orthographic knowledge and can use it when processing demands are relaxed. There do seem to be small ability differences in simple letter (bigram) processing, but these seem insufficient to account for higher level lexical effects. Evidence here is scanty, and this remains an open question.

Lexical Access and Context. Whatever the source of the lexical access problem, the evidence is that low-ability readers are adequate users of

contexts. Their time to name words is greatly facilitated in helpful contexts and inhibited by anomalous contexts. The fact that they are more
influenced by context is explainable by their slower basic word-identification rate. High-ability readers can be just as affected by context when
their basic word-identification processes are slowed down. These facts
again demonstrate the importance of retrieving linguistic information from
an *inactive* memory. Efficiency at this process is the hallmark of the high-
ability reader.

Remaining questions

It seems clear that a major unanswered question concerns the *exact* source
of ability differences in lexical access and decoding. Our data suggest
processes at the orthographic structure level that may include phonetic
coding. However, there are other possibilities. A major question, in general, is the role of phonetic processes in ability differences.

A second question is the extent to which the research on this young
population can be applied to other cases. The discussion needs to be extended to older readers, including adults, and to the subject of dyslexia.
These extensions are the topics of the next two chapters.

NOTES

1. Carpenter and Just (1975) have shown that in general this two-parameter
    model can be collapsed to a single-parameter model, in which the num-
    ber of mismatches is the parameter regardless of whether they arise from
    negation or from basic mismatches. However, negation does not seem to
    be psychologically identical to basic memory mismatches. Moreover in
    these data N and M do not seem to make the same contribution.
2. The data for four low-ability readers and two high-ability readers were
    better described by a falsification model, which assumes that any false
    sentence requires extra time. It predicts $FN > TN$. The falsification model
    is not the recoding model described for picture verification (Clark & Chase,
    1972; Trabasso et al., 1971), although it makes a similar prediction se-
    quence. The recoding model applies only to binary choices, so that *not
    above* can be recoded to *below*. This cannot easily apply to false nega-
    tives such as *an apple is not a fruit*. Instead falsification assumes a neg-
    ative encoding parameter N and a falsification parameter F that applies
    only to *final* mismatches. In effect it assumes that a *TN* sentence such
    as *an apple is not a sport* is "recoded" into *an apple is a fruit* (costing
    time N) and then verified with no extra cost. Thus $TN = t + N$. False
    sentences require an extra process of determining that the proposition is
    false.
3. Of course, there are other possibilities. For example, insecure semantic
    knowledge may be more of a problem in a false sentence. Thus a true
    sentence truly *verifies* what the reader thinks he knows, while a false
    affirmative may somehow challenge the reader's knowledge. However,
    error rates were low and equal for these two types of sentences.
4. This experiment and the next one are unpublished, but the data were

presented at the 1978 meeting of the Psychonomic Society by Perfetti, Riley, and Greeno (1978).

5. The partition of the reading time components reflected in the equations ignores some important factors—number of propositions, new words, verb complexity, etc. It does, however, demonstrate the role of some kind of integration work between sentences that is independent of computation.

6. IQ was measured during the previous school year by school-administered Otis-Lennon tests. These tests do have a significant verbal component. All subjects in the subgroups were average or above average on this measure.

7. Of course there are many other possibilities, since the main data are that intercept differences are large and rate differences seem to be smaller and more variable. One possibility for the rate variability is that less able subjects use highly variable search *strategies*, sometimes scanning right to left, or rechecking certain words, etc. For the large intercept differences there are also many possibilities in response execution, display orientation, etc. However, the fact that intercept differences were not large in the word task counts against these interpretations.

8. This interpretation of the linear components of the search data is consistent with analyses of variance in which performance with pseudowords vs. performance with words was a variable for each subject along with the other variables of the experiment. That is, there was a significant pseudoword and ability interaction and a significant three-way interaction of these factors with set size.

9. The average percentile rank on the MAT reading test was 74 for the high-ability readers and 24 for the low-ability readers.

10. At first glance, not finding a skill effect in forward search seems at odds with the previous search experiments. However, there is an important difference in the two cases. In these letter search experiments, the six letters extended for about 1.5 cm. In the previous search experiments, the items in the display were separated to encourage linguistic coding. Thus the most appropriate comparison may be with the bigram search experiment for set size 1. There the target was two letters and the display was two letters, about 1 cm. in width. It is in these conditions, as in the forward letter search, that relatively small differences were found. As in the bigram search, the ability differences in forward letter search were smaller for positive trials than for negative trials (but not significant in either).

11. By "perceptual memory" I am not referring to the sensory storage system identified by Sperling (1960), which lasts less than .5 sec under ideal circumstances. Rather, it is the kind of active visual synthesis or visual image retention that can be useful for a brief period (Posner, 1969). Its exact duration for letters may be under some active control (Kroll et al., 1970), but in the conditions of this experiment it is probably more than .5 sec and less than 4 sec, the two intervals used.

12. As evidence that context effects occur without consciousness, De Groot (1983) shows that priming effects in lexical access occur at subthreshold levels. Similar demonstrations come from Marcel (1983) and Fowler, Wolford, Slade, and Tassinary (1981). De Groot (1983) and De Groot, Thomassen, and Hudson (1982) also produce evidence showing that priming effects in lexical decision are weaker than commonly supposed,

restricted to words strongly and directly related to the prime. If this conclusion is correct, then we need to reconsider how context effects might
work in reading. One serious possibility, as we suggested in Chapter 2,
is that *automatic* context effects occur as part of lexical access. In terms
of lexical decision experiments, these would be "postlexical" processes.

13. Neely (1977) used single-word prime lexical decision tasks with adults.
It is not obvious what the time course for prediction is with discourse.
It may be quite a bit longer than the 400 msec required by single-word
primes. In discourse, the subject, in order to predict, must retrieve much
more information. This probably takes longer. My guess is that automatic activation and prediction may both play a part in our experimental situation.

14. Seegers and Feenstra (1982) report an experiment with Dutch children
which only partly supports this generalization. They found that third-
grade low-ability subjects showed more facilitation and inhibition than
third-grade high-ability subjects. For eighth-grade subjects there was no
difference in these effects and indeed no inhibition effects at all.

CHAPTER EIGHT

The Development of Reading Ability

This chapter addresses the important issue of the development of skill in reading. Our developmental scope is from elementary school reading beyond the first grade through adult reading. (*Beginning* reading is considered in Chapter 10.) The general questions are these: To what extent can the major sources of reading ability identified in verbal efficiency theory be extended to older readers? To what extent do ability differences among older readers require fundamentally different understanding?

THE SCOPE OF DEVELOPMENT

There is a wide range of reading talent even before a child learns how to read. This suggestion is a reflection of the central role that general cognitive and linguistic processes play in reading ability. Some general cognitive contributions to reading comprehension, as we have seen, are fundamental but not always illuminating. For example, specific knowledge is critical to reading but may play a limited role in explaining *generalized* reading ability. On the other hand, inference making is a fundamental cognitive ability that is generalized to the case of reading ability. Generally speaking, as basic cognitive abilities develop, they become more important for all cognitive activities, including reading. Nevertheless, we shall try to go as far as possible without paying too much attention to such abilities. We want to understand reading ability while assuming that other cognitive abilities are either equal or irrelevant for a particlar task.

A slightly different approach is needed for linguistic abilities. We do not need to assume that linguistic abilities are fundamentally different from other cognitive abilities. Rather, we assume that reading is patently a linguistic activity, no matter what other cognitive abilities are required by a particular reading task. The development of reading ability is partly included in the development of linguistic ability. This, unfortunately, does not mean that we will be able to explain the development of linguistic ability (let alone explain how it maps onto reading at each point). The point is that we will need to acknowledge the possible contribution of a range of linguistic abilities to the development of reading, even without a detailed understanding of these abilities.

The starting point is our understanding of the ability differences among elementary school children. These differences prominently include lexical access and verbal memory. The theoretical link between these ability differences is the efficiency of linguistic symbol activation and manipulation. There remains some question about how elementary this symbol activation process is. For example, there is some reason to believe that symbol *structure* (e.g., orthography) may be especially important, but it seems that simple symbol activation is also important. Is elementary symbol activation the fundamental factor? Similarly it seems that semantic processes are important ability factors, but it is possible that they can be accounted for, at a deeper level, by the same symbol activation process. The developmental perspective helps with such questions. It may be that the acquisition of skill in reading changes the relative contribution of these factors.

Development of Simple Verbal Processing Efficiency

In learning to read, a child develops an increasing ability to process printed linguistic symbols. As we have emphasized, this is not only the ability to recognize words but the ability to process them efficiently. We have also seen that decoding, letter recognition, lexical access, and semantic access are all important verbal processes in reading. These processes we group together under the category of *simple verbal processes*. *Decoding* is transforming a printed input into a speech form. *Letter recognition* is activating a printed letter symbol; it plays a role both in *decoding* and *lexical access*. One developmental issue, in part, is whether development in letter recognition accounts for development in decoding and lexical access. Another developmental question is whether any of these processes are accounted for by simple *symbol activation*. Symbol activation is simply the access and retrieval from memory of a symbol, regardless of its linguistic status. As a general processing constraint, the speed of *symbol activation* sets the limit on activation of particular symbol types, i.e., letters, higher level letter patterns, words, names, and word meanings.

Development in the Elementary School Years

A few studies have data on ability differences across several different age groups, although it has been typical to compare ability groups at two age levels. In one thorough, multitask experiment, Curtis (1980) tested second, third, and fifth graders and compared third- and fifth-grade readers of high and low ability. Tasks included word vocalization (naming), letter naming, word matching, and pseudoword matching. Within each grade level, high-ability readers performed better on these tasks. However, the best discriminators of high- and low-ability readers were word- and pseudoword-naming tasks, not letter tasks. These results agree with what we find generally for elementary school children in confirming the importance of decoding. Curtis examined the patterns of task correlations to assess whether the contributions of particular tasks to overall reading ability changed between the third and fifth grade. She found that a standardized listening comprehension test contributed unique variance to the reading comprehension measure. That is, its correlation with reading comprehension was significant even after its correlation with the symbol activation measures was accounted for.[1] The interesting age comparison is that listening comprehension accounted for more unique variance among fifth-grade readers than among third-grade readers. Correspondingly, decoding factors, especially word and pseudoword naming, accounted for less unique variance among fifth graders than among third graders. At the same time, there remained large differences in decoding between ability groups, even in the fifth grade.

The way to interpret such results is as follows: Decoding factors, not general symbol activation, are important ability factors throughout this period between second and fifth grade. In the second and third grades, however, ability differences in decoding may not be closely associated with differences in oral language processing. The two are correlated, but, for example, there are children who are weak only in decoding and not in listening comprehension, as well as children who are weak in both. As children get somewhat older, those whose earlier problem was only in decoding have become better at it. Those whose problem was both decoding and listening comprehension have not. Thus decoding still contributes high correlations with reading comprehension, but its *unique* independent contribution is less.[2]

The development of context-free decoding

According to verbal efficiency theory, context-free word recognition is the most salient characteristic of reading ability. An important development in reading ought to be a decrease in dependence on context. Stanovich, West, and Feeman (1981) provide a demonstration that such development can take place fairly early. Second-grade subjects practiced oral reading of isolated words over 15 consecutive days. Following this period of practice, subjects were given a word naming test in which the context was varied. Sentences were constructed to make the target more predictable and less predictable. One important result of this study is that the effect of practice on reading the target words was to reduce the effects of context. Furthermore, when subjects were tested some six months later, context effects were reduced even for unpracticed words. Thus, it is possible that during this six-month period the children were becoming more skilled at context-free word decoding in general. Whatever the underlying processes, there is an important skill development in which word identification gradually becomes independent of context. As we have seen, this development seems to come sooner for high-ability readers.

Symbol activation: What develops?
Of course, the developmental changes in the factors that contribute to differences in reading ability in part reflect what develops in reading processes generally. Generally, we assume that the simpler the process, the more quickly it reaches its maximum processing efficiency for an individual. That is, individuals will differ on their absolute performance rate for any symbol activation process, but each process may achieve its maximum development in the same order across individuals. Children will develop both letter- and word-recognition abilities.

A comprehensive study by Doehring (1976) provides some general developmental data on some of these processes. Doehring (1976) compared groups of children from kindergarten through high school on four different types of task: visual matching to sample, visual matching to oral sample, oral reading, and visual scanning. Within each task there were several variations in the materials. For example, numbers, three-letter words, four-letter words, three-letter (CVC) syllables, and three-letter consonant strings were among the materials for visual matching to sample. In this task, a visual target was displayed and the subject found a match in a three-item display set. Thus it is a visual search task. Doehring found that the overall response speed to all printed symbols decreased up to sixth grade (age 11). In visual scanning—essentially visual search through rows of symbols printed on paper—speed of search decreased beyond the age of 12, for all symbols. Searching for word targets never reached the rate of search for letter targets. A similar result for overall speed was obtained in the oral matching task, in which the subject heard a letter or word, etc., and had to find it in a three-item array. Response speed reached maximum by seventh grade (age 12). In naming, response speed continued to increase beyond the seventh grade, even for colors. However, by grade three, letter naming was as fast as color naming and word naming. Some of these naming data can be seen in Figure 8-1, which shows the increases in speed for four types of material—pictures, colors, letters, and words. Because these speeds were measured on oral reading from large chunks of material, their precision is not high and specific comparisons are not meaningful. However they show two things dramatically. One is how quickly word reading catches up with other kinds of symbol naming. The other is the very gradual increase in speed that continues for all tasks through grade 11. This implies that speed of symbol activation can continue to increase, but only very slowly, with increasing development.

One important point about word recognition is how easy it becomes relative to other processes that are superficially similar. In Figure 8-1, words are named more rapidly than pictures by the end of second grade. In a same-different task, Gibson, Barron, and Garber (described in Gibson and Levin, 1975) found word pairs judged more rapidly than picture pairs in grades four and six but not in grade two. In our study of third-grade children, low-ability readers were significantly *slower* naming single words relative to single pictures (Perfetti, et al., 1978). Even high-ability readers were faster for words than for pictures only for one-syllable names. (In contrast with Doehring's studies, the picture names were the same as the word names in the Perfetti et al. study [1978], assuring a meaningful comparison.) In general, adults and skilled readers read words more quickly than they name pictures. Exactly when children begin to activate word symbols more rapidly than corresponding pictures is not important. The point is that for high-ability readers this change does occur early in the elementary school years.

Some simple verbal processes, however, do continue to develop beyond elementary school. For example, in Doehring's (1976) study, speed continued to increase in naming and visual search beyond the seventh grade. In real reading situations, the demands of the reading task certainly increase for older readers. These elementary verbal processes involving symbol activation occur over and over again as the reader reads a text. However it appears likely that the low-ability reader in high school, like the low-ability reader in elementary school, is not efficient at all these symbol activation processes.

High School Readers

Research on high school readers of different ability has been carried out by Frederiksen (1978a; 1978b; 1981). This research confirms the assumption that symbol activation processes continue to be inefficient for low-ability readers who are slightly older. It also adds some additional insights into the processing factors in ability.

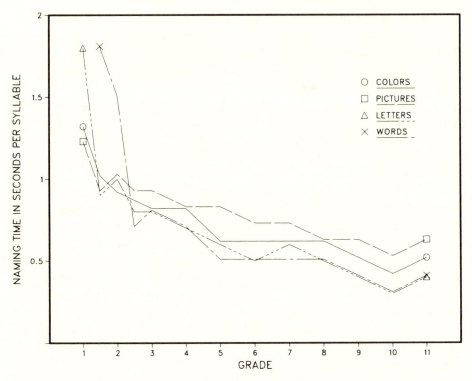

Figure 8-1. The development of naming speed for color names, pictures, letters, and words. Stimuli were presented as a series on a chart, not individually. The figure is constructed from data in Doehring (1976).

Frederiksen (1978a) designed tasks to tap three processing components of reading: perceptual, decoding, and lexical. The first two of these factors correspond to letter activation and word naming. The third, somewhat more complex, refers to activation and retrieval processes sensitive to specific lexical factors, including familiarity. The subjects for these studies were high school students with ability defined by the Nelson-Denny reading test. In earlier studies there were usually 20 subjects (Frederiksen, 1978a), and in later studies there were sometimes 44 subjects (Frederiksen, 1981). Subjects were grouped into quartiles of ability. The lowest quartile was below the 40th percentile; the other quartiles had percentile ranges of 41–85, 86–98, and 98–100. Thus only the lowest quartile would seem to be "low ability." Some of the results as summarized below.

Orthographic Structure. Subjects tried to recognize strings of four letters presented briefly under masking conditions. Recognition accuracy increased and decision latency decreased with the frequency of bigram patterns. Highly frequent bigrams, such as *TH*, were more accurate than low-frequency bigrams, such as *CK*. There was also a smaller effect of the positional frequency of the letters. The reading ability result is not quite what one might expect: Low-ability readers were more affected by bigram frequency than high-ability readers. This ability effect primarily separated the lowest group of five from the remaining subjects. The conclusion is that low-ability high school readers are relatively efficient at letter recognition in predictable (redundant) strings. High-ability readers are efficient in all letter strings.[3]

Decoding. Subjects vocalized single words and pseudowords. High-ability readers were faster than low-ability readers on both. Interestingly, the word latencies distinguished three ability groupings, with the top two quartiles indistinguishable. However the pseudoword latencies clearly distinguished four different reading levels. This is important evidence that, just as naming *accuracy* overestimates decoding ability of younger children, *word*-naming speed overestimates decoding ability of high school students (and probably adults). The low-ability readers were also more affected by words and pseudowords that began with longer initial consonant clusters, by two-syllable words (compared with one-syllable words), and by each additional letter to be identified.

Lexical Form. Subjects performed a lexical decision task in which letter strings were either all in capital letters (or all in lowercase) or they were in a mix of capital and lowercase (e.g., *WoRd*). The effects of case mixing were much greater for low-ability readers, especially for the two lower quartiles. The low-ability readers were also more affected by frequency. Frederiksen's (1978a) interpretation is that case mixing reduces visual access to familiar word forms and causes a greater reliance on decoding processes. This result at least suggests that low-ability readers depend more on visually familiar word forms.

Conclusions. The data summarized above were used to construct structural models of reading ability in Frederiksen (1978b).[4] One major processing factor contributing to general reading ability was multiple letter encoding. In our terms, this may refer to the rate of single-letter activation in the absence of any facilitation, e.g., the rate of encoding each letter in a poorly structured string such as *HTSR*. Interestingly, this factor seems to include a factor we have referred to as inactive memory. It includes, in Frederiksen's task, the difference in a name-level letter matching between *Aa* (same) and *Ad* (different). The different-name case takes longer. Different letter names cause a "mismatch" in activation, whereas same names are facilitated because the first letter primes the second. Thus a very general factor may be identified, roughly the speed of letter activation without facilitating context.

This does not mean that high-ability readers are less good at using orthographic structure. In another experiment Frederiksen and Adams (Frederiksen, 1981) varied the exposure duration of four-letter anagrams (e.g., *rtnu*) and pseudowords (e.g., *etma*). This is clearly a contrast between well structured and poorly structured letter strings. Here the critical measure was the increase in the number of letters a subject could report as the exposure duration was increased from 6 to 50 milliseconds. High-ability readers showed a larger advantage of structure than did low-ability readers. That is, as exposure duration increased, the rate of increased letter recognition was higher for pseudowords than for anagrams, showing the effect of structure. The highest ability group showed this effect more than the three other groups.

The best overall conclusion may be this: High-ability high school readers are superior in context-free letter activation rate; low-ability readers may be somewhat more dependent on orthographic context, but are not better at using it. At the very highest level of ability (above the 98th percentile) readers may take special advantage of structure, although they are not dependent on it.

Finally, there are also important decoding factors in the range of high school ability. The high-ability high school readers of Frederiksen's experiments were consistently faster at pseudoword naming than low-ability readers. This decoding factor is not completely dependent on the speed of letter activation from letter strings, although the two factors are correlated. Low-ability readers were especially affected by increases in pseudoword complexity, either by length or by vowel complexity. Thus there remain two distinct processing factors, interrelated but not reducible to a single process. One is rapid activation of letters, the other is the rapid activation of speech processes from pronounceable letter strings. That is, the two main ability factors among high school readers seem to be letter activation and decoding.

Adult Readers

It is not obvious that among adult readers the ability factors of childhood continue to be important. Adolescents and young adults have had increasing opportunity to master the simple verbal processes that are important in children's reading ability. However, there is reason to believe that simple verbal processes continue to be important in explaining ability.

There is one oddity concerning adult reading. Most of what we know about the simple processes of adult reading is based on research on college students. There have been important studies of adult reading in noncollege populations as well (see Sticht, 1977, 1979). For adult populations of low literacy levels the problems of ability are complex, although they undoubtedly include contributions of simple verbal processes. However, for what we know about adult abilities in these processes, we are largely dependent on college student samples. For such a population, we expect a selection process to have occurred, since children who demonstrated low ability in elementary school and high school are less likely to attend college.

The studies of Jackson and McClelland (1979) are informative for ability factors in adult reading. They defined reading ability by performance of college students on passages designed especially for the research. This performance measure was effective reading speed, which takes account of both accuracy and speed of reading: Effective reading speed = comprehension accuracy × reading speed. The comprehension measure is based on a short-answer test following the subject's reading of the passage. There was also a listening comprehension test based on this passage. The subjects of Jackson and McClelland (1979) were divided into groups of high vs. low ability relative to college freshmen and sophomores at the University of California, San Diego. High-ability readers were in the top quartile of effective reading speed, and low-ability readers were in the bottom quartile.

There were several tasks in the Jackson and McClelland (1979) study. One task that showed no ability difference was single-letter recognition threshold. Another no-difference task was thresholds for recognition of separated letter pairs. These results agree with earlier results by Jackson and McClelland (1975). Ability differences are not related to single-letter recognition accuracy.

Ability differences were found in matching tasks, which required the subject to decide whether pairs of visually presented stimuli were the same or different. The letter task required a "same" judgment to pairs of letters with the same name *(Aa)*; some pairs were also physically identical *(AA)*. The synonym task required decisions about whether two words had the same meaning. The homonym task required "same" responses to words that were homonyms *(doe, dough)*. The pattern task required judgments for pairs of nonlinguistic stimuli (□+, □□, ++, +□). The results for these tasks were that high-ability readers were faster than low-ability readers on all of them, including the pattern task. Ability differences tended to increase from the pattern task, which was fastest, to the homonym task, which was slowest.

There were two other tasks. One was a multiple-letter search task, with a display containing two, four, or six letters. (This task is similar to the one described in Chapter 7). The other was an auditory short-term memory task, in which the subject had to recall strings of consonants, as in a digit-span task. On the search task, high-ability readers had shorter times, but differences were in intercept not in slope. That is the *rate* of search was not different, but some component of general reaction time was. The auditory memory task produced very small but significant differences in mean number of items recalled, for example 5.64 versus 5.04 when strings were six items long.

Jackson and McClelland carried out further analysis based on intertask correlations, and the results of regression analysis support these conclusions: The best independent predictor of effective reading speed was listening comprehension measured on the same paragraphs. With listening comprehension controlled matching speed also predicted effective reading speed. The best matching task was letter matching. Jackson and McClelland interpret this as a name retrieval factor because all the matching stimuli were nameable and because a second experiment found no ability differences in speed of matching dot patterns. Thus this factor refers to a process of name activation in memory, not a simple perceptual process. Later experiments by Jackson (1980) suggest that this is a general visual access factor, however, one that depends on name codes in memory but *not* on alphabetic inputs. Jackson (1980) found that reading ability was related to the speed of matching categories of drawings as well as letter match. Thus, in terms of simple verbal processes, we conclude that there is a simple process of symbol activation that is important in college-level reading ability.

A decoding factor, independent of simple symbol activation, may be less important in this population. However, two results of Jackson and McClelland (1979) are interesting. First, high-ability readers were not only faster at the homonym task *(doe-dough)*, they were more accurate. Indeed, there was a significant correlation of accuracy with effective reading speed even after all other variables had been controlled. This may indicate specific word knowledge differences or underlying differences in knowledge of orthographic rules. In a second experiment, the speed difference in homonym decisions was found in pseudowords. Jackson and McClelland suggest that such differences could be accounted for by letter-encoding differences. Although their conclusion seems reasonable, with other, more demanding tasks of decoding it is conceivable that a different conclusion would be warranted. Recent research by Hammond (1984) suggests this is the case. Hammond's college subjects were comparable to Jackson and McClelland's, but unlike the latter, they performed pseudoword- and word-naming tasks in addition to letter matching. Vocalization latency in these tasks not only distinguished high-ability readers from low-ability readers; its correlation with ability remained significant even after controlling for the correlation between letter matching and ability. Thus, it appears that a lexical access factor may contribute to adult reading ability beyond simple symbol activation, provided a reasonably demanding access task is used. This vocalization latency measure is one of the most discriminating measures for children and adults.

Finally, it should be emphasized that the best predictor of reading effectiveness in Jackson and McClelland's study was listening comprehension. Listening comprehension was also a good predictor in Hammond's study. It is important to keep in mind that, whatever the contribution of lexical access or decoding skills beyond symbol activation, the contribution of a general language skill looms large—and at this point relatively unanalyzed.

Text reading

Much of the evidence on reading ability comes from tasks that do not require the actual reading of texts. In the long run, it is critical to learn whether the basic verbal processing factors that seem important in ability are observable when actual text reading is required.

A study by Graesser, Hoffman, and Clark (1980) illustrates one approach. They had college students read texts that varied in a number of ways. Some of the variables could be considered as higher text-level variables: (1) the extent to which the text was narrative (as rated by subjects), (2) the familiarity of the text, and (3) the relative number of new nouns (new arguments or new concepts). Others were more local text factors: (1) the number of words in a sentence, (2) the number of propositions in a sentence, and (3) the syntactic predictability of the sentence. Graesser et al. (1980) measured the reading times for a total of 275 sentences in 12 different texts. They found that the best predictor of reading times was the rated narrativity of a text. That is, sentences in texts that were more narrative-like were read more quickly than those in less narrative-like texts and this factor accounted for the major portion of systematic reading time variance.

However, a different picture emerged when Graesser et al. compared the faster readers with the slower readers. Only local text-processing factors differentiated the performance of fast and slow readers. Slower readers were more affected by the number of words and propositions and by syntactic predictability than were high-ability readers. Ability differences were unaffected by narrativity, the factor that was important for overall reading time.

These data thus confirm the importance of lower level processing components in adult reading ability for actual texts, provided that we can equate reading speed with reading ability. They also provide an important reminder for how general processing factors and individual differences can be related: It is not necessarily the case that those factors that contribute most to general processing are the same ones that produce the most important individual differences.

Reading Ability or Verbal Intelligence?

It is likely that at the college level what we have referred to as reading ability is not all that different from verbal intelligence. Nor should it be otherwise. The earnest effort to consider intelligence apart from reading ability has a practical purpose in the study of developmental dyslexia or of children's reading ability generally. The question with children is whether they should be expected to read better than they do read. But at the college level, even the "low-ability" readers of Jackson and Mc-Clelland are unskilled only relative to a college population; we should not be surprised if they score low on verbal tests of the Scholastic Aptitude Test (SAT). Nevertheless, even aside from matters of selection and other population characteristics, it is reasonable to assume that the verbal abilities that matter for verbal intelligence are some of the same ones that matter for reading, and vice versa.

In fact, studies of adult verbal intelligence suggest a shared view with adult reading. The studies of Hunt (1978; Hunt, Frost, & Lunneborg, 1973; Hunt, Lunneborg, & Lewis, 1975) have emphasized this same process that we have referred to as speed of symbol activation. The letter-matching task first described by Posner and Mitchell (1967) is the task that typically represents this process and it is the same task used in the Jackson and McClelland research. The ability measure on the letter-matching task is based on the difference in speed of two matching tasks. In a physical match, the physical identity (PI) of a pair of letters is the basis for a decision (e.g., *AA*). In a name-level match, the name identity (NI) of the letters is the basis for a decision regardless of their physical shapes (e.g. *Aa*). Hunt (1978) summarized several studies relating letter match times to the verbal ability of college students. There are consistent (but small) correlations between verbal ability and the difference between name matches and physical matches (NI $-$ PI). The interpretation of the NI $-$ PI difference is that it estimates the time to activate and compare single-symbol names while controlling for a visual matching component. It is a name retrieval measure. From the data summarized by Hunt (1978), it appears not only that NI $-$ PI correlates with verbal intelligence of college students but also that the NI $-$ PI difference may decrease with increasing absolute ability. For example, in a University of Washington study, non-University adults show NI $-$ PI differences of 110 milliseconds compared with a difference of 64 milliseconds for high-verbal University students. ("High-verbal" means high on a test similar to a verbal SAT). At the other extreme mildly retarded children show an NI $-$

PI difference of 310 milliseconds. As Carroll (1980) points out, most studies also find small PI differences related to ability. However, the critical result is that some name-level activation and comparison process beyond shape comparisons is implicated in verbal ability. As we have seen, it is also implicated in college reading ability. Carroll (1980) actually estimated NI − PI from the Jackson and McClelland experiments described previously. Its correlation with verbal SAT scores was nearly as high as its correlation with effective reading speed (r = .57)[5] (It did not correlate with listening comprehension.) Conclusion: Reading ability and verbal ability at the college level are related to the same simple verbal process, the speed of symbol name activation and retrieval.

It is interesting to consider the implications of the assumption that verbal ability depends on symbol activation and retrieval. The symbols that are accessed are essentially overlearned codes. Reading is the principal means for acquiring overlearned codes. Even the lowest ability college readers must have overlearned letter names to their maximum. What is the difference in the amount of "practice" with a letter symbol between a 9-year-old third-grade student and a 19-year-old college student? Assuming a conservative estimate of 1,000 words per day, the 19-year-old has accessed 3.6 million more words than the 9-year-old—at least 13 million more letters. That should be enough practice to give even the low-ability college student an edge over even the high-ability third grader. In that light, it is surprising that activation of overlearned symbols has any variance among individuals, let alone any strong relation to general ability.

Another way of putting this is to note that differences among adults in such things as vocabulary are certainly more striking than differences in NI − PI. However, vocabulary differences continue to be accompanied by differences in simple verbal processes. For example, Butler and Haines (1979) found that college students of high vocabulary had faster decoding times than low-vocabulary students. Of course, this does not mean that decoding speed is more fundamental than vocabulary. Decoding itself may reflect speed of general symbol activation. Although cause and effect cannot be determined, it does seem safe to conclude that simple speed of symbol activation has reached its maximum efficiency by college age. Differences among individuals have had years to emerge in all areas of verbal processing, including word knowledge.

Finally, we consider speed of semantic access. This process may be part of the ability differences among children. But semantic access seems a less likely explanation for adult differences to the extent that simple symbol activation and retrieval can account for other aspects of verbal processing speed. An experiment by Goldberg, Schwartz, and Steward (1977) provides a comparison between decoding and semantic access. Subjects made matching decisions based on name identity *(Deer-dear)* and semantic category identity (deer-elk), as well as on physical identity (deer-deer). High-verbal subjects were faster than low-ability subjects on the name task (363 milliseconds) and the semantic category task (360 milliseconds). (The difference was only 136 milliseconds on the physical match task.) Thus, these data suggest that semantic access speed differences can be accounted for by decoding level differences.

The semantic access factor is also seen in a study in which subjects performed several picture- and word-matching tasks (Hunt, Davidson, & Lansman, 1981). For example in semantic categorization, subjects decided whether a picture or a word was an instance of a prespecified semantic category, a task earlier used by Hogaboam and Pellegrino (1978) in a study of verbal ability. Subjects also performed word-matching tasks (DATE-date vs. DATE-gate), a semantic matching task (ELK-DEER), a simple-choice reaction task, and some paper-and-pencil tasks. Performance was correlated with individual scores on the Nelson-Denny reading test, including its vocabulary test. The magnitude of the correlations was around .3 for all semantic categorization tasks, both for pictures and for words. The correlation between word and picture categorization was $r = .99$! Clearly we do not need a specific print-decoding component for these subjets. Hunt et al. (1981) also examined their processing tasks for a common component, one that they refer to as speed of memory access. The results of a factor analysis indicated that these various processing-speed tasks included a common factor that accounts for 75% of the task variance.[6] Since none of these tasks involved stimuli as simple as single letters or digits, this factor may not be simple symbol activation. However, the fact that the tasks included both name-level and semantic-level information (and pictures and words) makes it possible that such is the case. Furthermore the size of the correlations is consistent with other studies using letter name (Aa) matching (Hunt, 1978).

One other result of Hunt et al. (1981) is interesting. There was, in addition to the processing-speed tasks, one task that did not involve speed nor significant use of words in memory. It was a paper-and-pencil test in which subjects verified sentences describing visual symbol arrangement. The symbols were either ($\frac{*}{+}$) or ($\frac{+}{*}$). and subjects indicated *true* or *false* to sentences which varied in linguistic complexity, e.g., *Star isn't above plus* (Clark & Chase, 1972). Performance on this task seems to test syntactic processing and manipulation of symbols in memory more than speed of symbol access. High-ability subjects outperformed low-ability subjects, and the correlation with ability was independent of the matching-speed contribution to ability.

To summarize this semantic access issue: Although the experiments of Hunt et al. (1981) do not directly test whether there is a semantic access speed difference beyond general decoding or symbol activation, the fact that there was a very general factor in all the tasks indicates there is not. The results of Goldberg et al. (1977) are in agreement with this conclusion. Our conclusion is this: Younger children of low reading ability may have semantic information less tightly bound to a memory symbol, so that activation of the symbol is not always sufficient for the semantic information to become useful. However, with increasing verbal experience, common semantic category information becomes tightly bound to the symbol-name. That is, semantic information is specifically connected to the symbol-name rather than conditionally connected to the symbol-name. High-ability adults differ from low-ability adults in the basic speed of access and retrieval of this symbol-name but not in whether symbols are bound to their semantic values. One qualification: We do not cease to learn new words at any particular point in development. If a college student has a fragile knowledge of a particular word, e.g., *predicate* or *Renaissance*, the semantic information may not be bound to the name, but may be very context dependent.

Rate-limiting Processes in Development

The key features in development of reading ability can be summarized. First, there are simple verbal processes that affect the linguistic processes of reading: (1) general symbol activation and retrieval; (2) letter recognition; (3) word decoding; (4) semantic access. These processes are related to verbal ability (i.e., reading ability or verbal intelligence). Second, there are refinements to these processes that occur in individual development. There is reason to believe that ability differences among children include word decoding in addition to letter recognition and perhaps semantic access in addition to decoding. It is likely that for children above the second grade, letter recognition is not an additional ability factor beyond general symbol activation. After all, two years of reading instruction, even if partially unsuccessful, ought to bring simple letter codes into memory up to the level of any other symbol. (It is hard to imagine more overlearned symbolic codes—certainly not digits or pictures.) However, by the time readers become adults, things have changed. Semantic access and decoding have reached maximum efficiency independent of general symbol activation.

The developmental question here is what process sets overall processing rate limits. General symbol activation and retrieval set the overall limits. Nothing can occur faster than the access and retrieval of an over learned symbol name. However, until this limit is reached, there are opportunities for other processes to be rate limiting: first letters, then word names (decoding), and to some extent, semantic processes. Of these, according to the evidence, the most important is decoding. Throughout

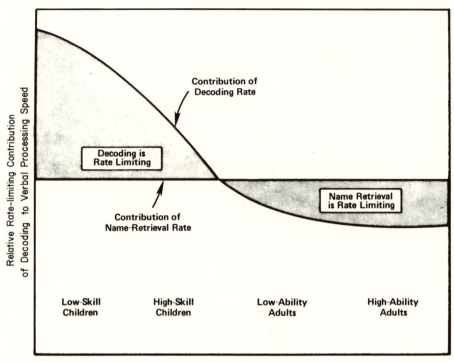

Level of Reading Skill Development

Figure 8-2. Schematic representation of the relative contribution of decoding and name-retrieval rates to verbal processing with increasing reading skill. At low levels of skill, decoding is rate limiting. At higher levels, name retrieval is rate limiting.

much of the elementary school period, decoding ability is rate limiting.

We can represent the general developmental pattern by considering just decoding and general symbol activation, as in Figure 8-2, which shows, hypothetically, the contribution of decoding to verbal processing speed, relative to general symbol-name access and retrieval. The latter is represented as making an invariant contribution. This does not mean that access and retrieval of symbol names does not increase in efficiency, only that these processes make a relatively constant contribution to overall ability differences in verbal processing speed. For young children and especially for low-ability readers, however, access-retrieval is not the rate-limiting process; decoding is. As decoding improves in efficiency, it is no longer rate limiting. Individual differences remain profound, but they derive increasingly from basic limits in symbol name access and retrieval—a process that operates on linguistic and nonlinguistic symbols in memory. While Figure 8-2 shows only decoding, similar representations are plausible for letter recognition and semantic access. At least letter recognition, however, would very quickly fall below the rate limit of general symbol name activation and retrieval.

Process-limited and Knowledge-limited Comprehension

It is one thing to point out that individual abilities in simple verbal processes are associated with individual abilities in reading and verbal intelligence. It is another thing to be certain that the abilities in simple verbal processes are causally connected to the general abilities. The general case for verbal efficiency has been made in earlier chapters. It is clear that efficiency differences in simple verbal memory processes can be causally related to comprehension. However, in the development of reading ability it is useful to distinguish two components to this relationship. One component is that simple verbal processes affect comprehension or other complex verbal performance because they are *process limiting*. The second is that simple verbal processes are *knowledge limiting*. The first component refers to the limitations on comprehension processes imposed at the time of processing. The second component has its effect in the history of the individual. The efficiency of simple verbal processes has directly or indirectly influenced the acquisition of knowledge, i.e., learning.

The development of reading ability thus includes two consequences of verbal efficiency: A given reading task is limited by the momentary efficiency of processing and by the previous learning of the individual. The amount of prior learning will have been partly influenced by simple verbal processes. But there is another important developmental principle in operation: The acquisition of knowledge is only weakly determined by verbal efficiency. For example, large differences in verbal efficiency should produce large differences in knowledge, other things equal. And small differences in verbal efficiency should produce small differences in knowledge, other things equal. However, there is no reason to assume that two individuals who differ by the slightest microsecond in symbol retrieval will differ in knowledge in one direction rather than another. *Beyond some theoretical limit in efficiency, knowledge acquisition is not limited by efficiency.* If anything, only the ease of acquiring knowledge is affected at the upper levels of efficiency. This is why correlations between name retrieval and college verbal ability are so modest. They would be much more profound among an unselected population, just as they are among children. The fact is that the acquisition of knowledge is subject to many other influences at the higher levels of efficiency.

Summary

The development of reading depends in part on acquisition of simple verbal processing ability. Since the processes may develop at different rates, ability differences among younger children may not be exactly the same as among adults. Nevertheless it is clear that certain verbal processes continue to be important through elementary school, high school, and adulthood. At all age levels, high-ability readers show more rapid access to symbol names in memory. In the elementary grades, decoding is a critical ability factor relatively independent of general symbol activation and retrieval. There is, however, an indication that decoding begins to make less of a unique contribution to overall ability as early as the fifth grade. Letter processes seem to reach a high level of skill as early as the second grade. By college age, the decoding process appears not to make a unique contribution to ability beyond its dependence on general symbol activation. Listening comprehension is an important factor throughout developmental levels.

THE DEVELOPMENT OF GENERAL LANGUAGE COMPREHENSION

There is much more to reading ability than efficiency at simple verbal processing. Indeed, as a general rule the more demanding the reading task, the more noticeable is the contribution of higher level language abilities. It is conceivable that higher level language abilities are partly dependent on lower level abilities, but even if they are it is important to understand their role in the development of reading ability.[7]

The important general language processes include those described in Chapters 3 and 5: the assembly and integration of propositions, inference processes, schema activation, and text modeling. In addition they include metacognitive skills and strategies. They also include explicit linguistic knowledge, which continues to develop. However, we have very little evidence about the development of differential abilities in these areas. Instead we have some interesting possibilities to consider.

Syntactic processes
It is quite likely that throughout the elementary school years and even beyond, there continues to be significant development of linguistic abilities. For example, syntactic processes that may be especially important for reading may require further development over the basic grammatical competence that young children have. Indeed, it is likely that reading places demands on syntactic abilities that are not placed by oral lan-

guage processing. Oral language is so heavily dependent on context that it seldom places heavy demands on syntactic processes. Reading, much less contextualized, seems to require more syntactic processing.

As an example of a syntactic process, consider anaphora. The reader or listener, when an anaphora is encountered, must decide where in his model of the text the antecedent is to be found. A pronoun, *he* or *it*, must refer to something already mentioned. It is common in texts for there to be more than one possible referent. Frederiksen (1981), in his studies of high school readers described earlier, had subjects say what the referent was for certain pronouns encountered in short texts. For example:

(1) *Modern advertising does not, as a rule, seek to demonstrate the superior quality of the product.*
(2) *It plays up to the desire of Americans to conform, to be like the Joneses.*

In sentence (2) *it* refers to the first noun phrase of sentence (1), modern advertising. Low-ability readers and high-ability readers handle this easily. But suppose a different sentence appears as (1a):

(1a) *The superior quality of the product is not, as a rule, what modern advertising seeks to demonstrate.*

Now the antecedent of the pronoun of sentence (2) is the second noun phrase of the preceding sentence. That is, the antecedent is still *modern advertising*, but it is no longer at the beginning of the sentence. Frederiksen's (1981) low-ability readers had some problems with this second case, taking quite a bit longer to decide on the correct referent. High-ability readers had no problem with it. This suggests that low-ability readers had a topicalization strategy: When the pronoun begins a sentence (i.e., is the topic of the sentence), look for its antecedent as the topic of the preceding sentence. The high-ability reader, more flexible perhaps because of greater efficiency, has no need for such a strategy.

We probably do not want to conclude that low-ability readers have poor syntactic strategies. This is just one kind of evidence that syntactic processes in reading can become important, especially as the reader needs to integrate propositions. We have no reason to suppose that the ability to determine anaphoric reference has not developed well before this high school age. What may be important is whether the demands of textual syntax can be easily met by the reader. Meanwhile, some syntactic abilities continue to develop along with reading ability. The low-ability reader may not only be less able to apply his syntactic knowledge to written text, he will also acquire less syntactic knowledge through reading.

Inferences

If a reader encounters a sentence such as (1), (2), or (3), we might expect him to make some inferences:

(1) *Fred dug a hole in his backyard.*
(2) *Mary knew that it was raining in Youngstown.*
(3) *Molly walked to the store carrying her umbrella on her arm.*

For (1) the reader might infer that Fred used a shovel; for (2) he might—should—infer that it *was* raining in Youngstown; in (3) he might infer it was not raining at all. These examples, because they are very ordinary sentences, illustrate the pervasiveness of inferences in comprehension. Texts are never fully explicit. They also demonstrate the variety of inferences: the one in (2) is logically impelled; those in (1) and (3) are not. Apparently, everyone does not make inferences in such cases. More important, there are developmental differences. Younger children (age 7) are less likely to make inferences spontaneously for sentence (1) compared with older children (age 9) (Paris & Lindaur, 1976). There may be some circumstances in which even adults do not make such inferences when they read (Vonk & Noordman, 1982). Nevertheless, we have a clear candidate for an important development in reading ability. High-ability readers may make inferences at a younger age than low-ability readers. We should not be surprised if at any given age high-ability readers make more inferences. We know they have more processing resources for inferences because they have a higher level of verbal efficiency. In general, we can make the following assumption for development. The contribution of inference making to overall reading comprehension will increase with development. Its contribution to ability differences may be expected to be high for children as text reading becomes more demanding. For very young children and for adults general inference making may be less of a factor except as it is controlled by specific knowledge.

Listening and reading

The relationship between oral and written language is a very close one. And we know that reading ability and listening ability are closely associated throughout the age range (Berger & Perfetti, 1977: Curtis, 1980; Jackson & McClelland, 1979: Perfetti & Goldman, 1976). However there are some differences between oral and written language processes. One of the most important is that printed texts are more explicit. A second is that oral texts are more pragmatic. Olson (1977) has made a case for the idea that our propositional idea of language is made possible only by literacy. Speech contains utterances, written texts contain propositions. Utterances at an abstract level, of course, are also propositions. However, because they occur in rich communication settings, and because they usually refer to a shared field of experience between speaker and listener, their *propositional* content only loosely determines the *communication* content. By contrast, in print the propositional content becomes dominant. The writer seldom has one particular reader in mind as the receiver of his message (personal letters are an interesting exception). The result is a more explicit text, fewer gaps, less inferring, and more attention required by the reader to the language *in* the text.

It is possible, even likely, that the development of reading comprehension ability is not a straightforward transfer of listening processes to print. Of course, there are numerous differences between printed and spoken language. Spoken language is more contextualized, more transitory, and does not require the learning of a new code. Printed language is more propositional, more permanent, and does require some code learning. There are other differences that are important but essentially derivable from one of these differences (see Chapter 1). However, the differences between speech and print may change with development in a number of ways. At first, decoding is a salient difference between speech and print. Another difference becomes noticeable later, as the propositional content and explicit style of written texts becomes salient. Additionally, it is quite likely that reading fosters some significant linguistic growth for the child. This is a consequence of the fact that print is relatively decontextualized. Since the printed text cannot rely on the communication context that speech can, it must direct attention to sentences as the means for obtaining meaning. It is possible to comprehend oral language with only primitive syntactic abilities, but it is difficult to understand written texts without syntactic abilities because they depend on explicit propositions contained in sentences. In short, we can assume that significant developmental events occur as a child attempts to transfer oral language strategies to written texts. And we can wonder whether these events contribute to the development of reading ability.

Hildyard and Olson (1982) carried out an interesting experiment that tests the hypothesis that oral and written texts are different in their explicitness. They presented spoken and written stories to third- and fifth-grade subjects who were either high or low in reading ability. They were interested in whether readers recalled more explicit information than listeners and whether this depended on age or reading ability. A story example is shown below, along with four types of information that were tested: structural vs. incidental, and implicit vs. explicit. Structural information consisted of data important to the central story content. The implicit information was not given in the text but was inferable.

Susan and Jonathan lived in a house in the middle of the city. At the end of their backyard there was a large maple tree. Susan and Jonathan often played under the maple tree in their sandbox. One morning they found something in the sand. It was tiny and white. Susan went into the house to find a container to put it in. She went up to her bedroom and came back carrying a black and white box. It's too big, said Jonathan.

So Susan found a handkerchief and some Kleenex. They put the handkerchief on the bottom of the box and laid the tissues on top of the handkerchief. Jonathan carefully laid the strange thing on the tissues.

Next morning their teacher was interested in what they had found and gave the whole class a lesson on how birds' eggs hatch.

1. *Structural implicit:*
   Susan and Jonathan found a bird egg.
2. *Structural explicit:*
   What they found was small and white.
3. *Incidental implicit:*
   The sand box was at the end of the backyard.
4. *Incidental explicit:*
   They lived in the city.

Following listening or reading, subjects took a two-choice test asking questions that depended on one of these four types of information. For example, a test item for structural implicit information was a choice between *Susan and Jonathan found a bird egg* and *Susan and Jonathan found a stone.* Children actually answered two questions: Which choice was better? And did the story really say this? The important results are these: (1) High-ability readers performed better than low-ability readers on both listening and reading. (2) Listening and reading differed on incidental information, with written texts producing better memory for incidental information. For structural information, reading and listening were equal. (3) On the strict question of whether the information really occurred in the story (as opposed to merely fitting in appropriately), fifth-grade subjects were better than third-grade subjects and high-ability readers were better than low-bility readers. (4) Finally, and perhaps most interesting, children who read were able to discriminate explicit from implicit statements better than children who listened, when response bias was accounted for. Fifth graders and high-ability readers were also better at discriminating explicit from implicit statements than were third graders and low-ability readers.

This experiment seems to confirm the hypothesis that written texts produce more explicit processing than oral texts. Note that this effect does not depend on reading ability nor on age. There is no evidence that low-ability readers fail to adopt an appropriate explicit text strategy in reading. An interesting developmental question may be this: At an advanced stage of development might we expect the listening-reading difference to disappear? That is, does listening become more like reading under some conditions?

Story structures

Yet another type of higher level knowledge is the child's knowledge of the structure of stories. We have already noted that the evidence for ability differences in appreciation of structural importance is inconclusive (Chapter 5). The Hildyard and Olson study just described did not find an ability interaction with structural versus incidental information. Young children master the simple structural elements of stories at an early age (Stein & Glenn, 1979), and even low-ability readers, at least among 12- and 13-year-olds, reflect structural elements in their story recalls.

Linguistic memory

We know that working memory continues to differentiate high- and low-ability readers throughout the range of development (see Chapter 5). Adult verbal ability is also related to auditory memory span (Hunt et al., 1975; Lyon, 1977). However, we have suggested that linguistic memory and simple verbal processes are related: They both manifest the manipulation of symbols in a memory system. We do not know whether they are exactly the same things, but they do demand the same recurring symbol manipulation process. There is, however, an additional issue for development of reading ability. More complex texts tend to make greater absolute demands on memory as the reader gets older. Unless verbal efficiency keeps up with these demands, the reader will experience memory difficulty.

Summary

There are many contributors to reading ability development that go beyond simple verbal processes. We have described only a few of these to indicate that processes of linguistic inference may continue to develop and play a role in comprehension ability. However, we do not know much about other language and speech factors that seem to be related to reading comprehension. Important differences between oral language and reading processes seem to exist, but not differentially for high- and low-ability readers. Similarly, structural knowledge of stories develops too early to have much impact in later development. The one critical factor common to reading and general language processing is linguistic memory. It is not necessarily independent of other symbol-processing factors but it may appear to be when it is observed in complex texts. More is known about the development of simple processes related to reading than about more complex general linguistic processes.

The development of reading ability from the elementary school years depends on the development of two general components. One is the efficiency of simple verbal processes. Children develop efficiency in letter recognition, decoding, name retrieval, and semantic access as their reading fluency develops. A general symbol activation rate may set the general limit on efficiency throughout the development of fluency. However, for children of low ability, decoding, rather than general symbol activation, may be the rate-limiting factor. The second general component is the development of higher level language comprehension ability. This component is not independent of simple verbal processes, because acquisition of the latter may depend in part on effective reading.

NOTES

1. This unique variance concept is the residual correlation between two measures following the removal of all partial correlations with other variables entered into a multiple regression analysis. In the Curtis study, each independent variable's unique contribution was determined by entering it last into the multiple regression.

2. Other interpretations are possible and all must be taken with the critical acknowledgement that they are limited by measurement issues. Correlations among test scores reflect the difficulty of the tests, in part. If the decoding tests do not increase in difficulty while the listening test does— or, equivalently in this case, if the listening test has a built-in increasing performance standard—this variance affects the correlations. The interpretation offered is only consistent with the Curtis data, not compelled by them.

3. This result seems to contradict the effect of orthographic structure reported in Chapter 7 and in results of Massaro and Taylor (1980). In both cases, high-ability readers showed more benefit of orthographic structure. However, there are essential task differences. In the experiments of Chapter 7, the task was *backward* search, presumably a memory effect. In Massaro and Taylor, orthographic effects were found in forward visual search, but from their data it appears that most of the high-ability readers' advantage came from real words. Furthermore, in a later experiment Frederiksen (1981) found that high-ability readers could use orthographic structure more effectively than low-ability readers in pseudoword recognition.

4. Frederiksen's (1978b) procedure used the structural equations procedure of Joreskog and Sorbom (1978). Alternative rational models were based on hypotheses about what processing components were present in each task.

5. In fact, correlations between general ability test scores and NI−PI are usually smaller than this, typically around .30. As some have pointed out, these small correlations should make us realize that there is much more to intelligence than speed of verbal processing (see Sternberg, 1981). On the other hand, it should be kept in mind that NI−PI provides essentially a partial correlation.

6. Hunt et al. (1981) used the results of this principal-components factor analysis to compute canonical correlations, i.e., correlations between verbal ability measures and the factor loadings of the processing tasks on the memory access factor. This confirmed the contribution of a single

memory access factor to Nelson-Denny reading scores, $r = .69$. As Hunt et al. (1981) point out, this correlation is an estimation of the *maximum* relationship between reading ability (verbal ability) and speed of memory access because canonical correlations are sensitive to random error in the data.

7. In some of the studies there has been a statistical independece between some general measure of comprehension and specific processing-speed tasks (Curtis, 1980; Hunt et. al., 1981; Jackson & McClelland, 1979). This does not necessarily demonstrate that the comprehension processes underlying the general comprehension test are independent of the verbal processes measured by the speed tasks. The reason is that the measures of general comprehension are very similar to the reading comprehension tests which provide the dependent measures. For example, Curtis (1980) and Jackson and McClelland (1979) both used global listening comprehension tasks that were mirror images of the criterion reading comprehension tests, whereas the processing-speed tasks were quite different. An informative experiment would be to include batteries of listening and reading comprehension tests for comparable text units.

CHAPTER 9

Dyslexia

So far there has been much to say about reading ability in general and the factors that contribute to low ability. But what about *dyslexia* or *specific reading disability?* These two terms are used interchangeably to describe the problems of a specific subpopulation of low-ability readers. In general, *dyslexia* occurs among readers of very low ability whose IQ scores fall into the normal range. The typical requirement for the "dyslexia" label is that a person be of normal intelligence, with no neurological or sociological basis for the reading disorder. For children, the degree of reading disability is measured normatively—at least two years below grade level. There are some problems with the definition of dyslexia. For example, Vellutino (1980) points out that the requirement of two years below grade level is more difficult to meet for a second-grade reader than a fifth-grade reader. And the IQ definition question is very problematical, because IQ measured on verbal items presumably reflects some of the same abilities tapped by reading measures. Thus it is typical, but far from universal, to apply the "normal IQ" criterion to nonverbal IQ scores.

In the definition of dyslexia used here, we will apply these criteria as follows: A dyslexic is a child who is normal or above at least in nonverbal IQ, two years behind in reading achievement, and with a reading disability that is not explainable primarily by social, economic, motivational, or emotional factors. (For further discussion of these definitional issues, see Duane, 1979; Gordon, 1983; Jansky, 1979.)

A central question is this: Are there fundamental differences between low-ability readers, as we have considered them up until now, and dyslexics? Or are some of the general features of low-ability readers also applicable to dyslexics? In other words, are the factors relating to verbal processing and decoding applicable to specific reading disability as well as to the sort of low-ability readers discussed in previous chapters? In fact, the differences in population description may be minor. We have applied the category of "low-ability readers" only to children of normal IQ, although in many cases they were less than two years behind their grade level in reading.

Our focus here will be on *developmental dyslexia*, that is, reading disability among children without brain trauma. There is another disability, *acquired dyslexia*, which is the result of brain injury. Cardiovascular accidents (strokes) are the most common causes of acquired dyslexia among adults, resulting in a number of differentiated reading disorders (see Benson, 1981, for a description of these dyslexias). We will refer to cases of acquired dyslexia only to illustrate certain specific implications for how reading processes may be organized and reorganized. It is possible that the neurological factors that are central to acquired dyslexia are also involved in developmental dyslexia. Certainly some of the theories of developmental dyslexia that we shall examine suggest that anomalies of cerebral organization are part of developmental dyslexia. And it is possible that these neurological factors are different from those of the stroke victim mainly in degree and detectability. Nevertheless, there are as yet no firm grounds for linking developmental and acquired dyslexia in any specific way.

In the case of developmental dyslexia, we want to consider especially the extent to which verbal processing factors are instrumental in reading disability. There is one difference, however, between specific reading disability and the more general reading ability issues we have considered until now. In the case of general reading ability we have tried to understand reading comprehension as well as word reading. Ability differences in lexical processes have been examined to explain differences in comprehension. In the case of dyslexia, there is less need to justify a focus on lexical processes. There are no disagreements about the significance of word-reading difficulty for dyslexics. Problems in reading words are the salient difficulty for such readers. Indeed the prototype dyslexic, of any age, is one who has specific difficulty reading words and the word "dyslexia" itself contains this idea of word difficulty. Thus the earliest clinical observations of dyslexia and the research on dyslexic children have essentially tried to explain word-reading difficulty. Before we consider the possible verbal basis for dyslexia, we will briefly review some of these explanations.

CONCEPTUALIZATIONS OF DYSLEXIA

The fact that there are children who seem perfectly normal in most respects except for not being able to read has caused much interest among scientists, medical people, and educators with very different perspectives. It is not surprising that there are very different theories of dyslexia and many different ways to describe it. This section will not be an exhaustive survey of different approaches, but it will focus on some of those that have received the most attention. These include the idea that dyslexia reflects a visual-spatial deficit, the possibility that dyslexia includes distinctive subtypes, and the assumption that there is a neurological basis (or perhaps two) for dyslexia. After discussing these possibilities, we will return to the factors most closely related to general reading ability, namely verbal and general linguistic factors.

Visual-Spatial Deficits

One traditional view of dyslexia is that it reflects disturbances of visual perception. Orton (1925, 1937) believed that the dyslexic individual has an anomalous cerebral organization, essentially a failure or delay in developing cerebral dominance. Orton linked this neurological hypothesis to his observations of dyslexic children, especially their tendency toward letter reversals in reading and writing as well as their tendency to show typical motor correlates of lack of cerebral dominance (e.g., lack of handedness). There have been many discussions of cerebral organization and dyslexia since Orton, and a search for neurological bases for dyslexia has been, understandably, an important part of the specific reading disability research. Here we will ignore for now the neurological hypothesis and focus on the observation most important to reading: that dyslexic children make letter reversals in reading.

There are two kinds of reversals. In one case, a word is read so that its letters are reversed, for example, *nip* is read as *pin*. In the other case, a letter is read as if it were a mirror-image reversal, for example, *p* for *b*, or *d* for *b*. In fact, *b*, *p*, and *d* are really the only letters that are reversible in most type fonts. Apparently, both kinds of reversal errors are made by beginning readers to some extent. However, they are not so common as might be expected, and the two reversal types seem to be independent of each other. The study of Liberman, Shankweiler, Orlando, Harris, and Berti (1971) found that there was virtually no correlation between the occurrence of a sequence error *(pin for nip)* and the occurrence of a letter orientation reversal *(dad for bad)*. Combined, such reversals accounted for only 25% of misread letters among low-ability second-grade readers. The reversed orientation errors, furthermore, were quite variable across and within low-ability readers. They do not seem to represent a consistent individual characteristic.

Another interesting fact about letter orientation reversals is that they are very rare when individual letters are presented at rapid exposures

(Liberman et al., 1971). They occur much more frequently in reading real words. This is a strong suggestion that reversals are part of lexical access rather than letter perception. If we think of lexical access as an activation process with links between letters and words, such errors are explainable without reference to spatial anomalies. For example, the letters *b-a-d* might activate *bad* but also *dad* and *mad*. If the activation process is slowed down, as it may be for a beginning reader or a dyslexic, some letters will dominate the activation process over others. For example, *d* may send out more early activation than *b*, and words having *d* anywhere in them achieve high activation. Thus, with some probability (low but above 0), *dad* will be read instead of *bad*. One test of this idea is that errors of *bad* → *dad* should be more common than *bin* → *din*. However, *bin* → *din* would occur more often than, for example *bin* → *fin*, because both *b* and *d* receive activation from the feature level. There could even be activation from the phonetic level as /b/ and /d/ share the feature of voicing. Whatever the source of such reading errors, there would be nothing special about spatial orientation by this analysis.

There have been many studies of spatial factors in dyslexia, and certainly the conclusions of many of them are consistent with a spatial hypothesis. On the other hand, evaluations of the evidence have often been either inconclusive (Benton, 1975) or decidedly negative (Vellutino, 1979; 1980).[1] Vellutino (1979) argues that errors that appear to be spatial in origin may result from verbal factors instead. Dyslexic children who make errors in reading letters (or words) are demonstrating failures to have verbal codes reliably associated with letter forms (or word forms). These codes could include letter names, phonemes, and word names.

One particular study from Vellutino demonstrates that spatial problems in dyslexics may not be serious. Hebrew words were presented visually to normal and "poor" readers between 7 and 12 years old (Vellutino, Smith, Steger, & Kaman, 1975). Subjects copied the words from memory following a brief exposure. Given that subjects were viewing an unfamiliar alphabet, errors were expected. The important result was that low-ability readers performed as well as normal readers in this copying-from-memory task. Furthermore, the two groups did not differ in their tendencies to make orientation and sequencing errors, contrary to what might be expected by a visual-spatial hypothesis. In fact, for Hebrew letters that could be considered spatial transformations of Roman letters (e.g., ٦ as a transformation of *L*), disabled readers made *fewer* orientation errors than did normal readers. The critical fact here may be that the unfamiliar Hebrew letter forms do not have verbal labels. In an earlier study in which English words were copied following brief presentation, disabled readers were equivalent to normal readers provided the words had only three or four letters (Vellutino, Steger, & Kandel, 1972). Such results, along with many others (e.g., Shankweiler & Liberman, 1972) direct us away from an understanding of dyslexia based primarily on visual-spatial factors.

Of course it is possible that a relatively small percentage of dyslexics have fundamental visual-spatial problems. Pavlidis (1981) reports that dyslexics show abnormal eye-tracking performance in a task that does not involve reading. Subjects merely follow with their eyes a horizontal display of lights in which each successive light cycles on and off in turn. Dyslexics were reported to make more saccades and more regressive eye movements than normal subjects. The possibility that dyslexics have abnormal eye movements in reading would not be surprising, and this problem has in fact been observed (Tinker, 1958; Olson, Kliegl, & Davidson, 1983a; Pirozzolo & Rayner, 1978). But such differences do not implicate a visual-spatial deficit as a cause of dyslexia since abnormal eye movement patterns would also be a likely *result* of reading disability. On the other hand, eye movement abnormalities in a nonreading eye-tracking task would suggest a causal role for a visual-spatial factor in dyslexia. However, there have been consistent failures to replicate the pattern of abnormal eye tracking observed by Pavlidis (1981). At least three different studies report no difference between normals and dyslexics on the eye-tracking task (Brown et al., 1983; Olson, Kliegl, & Davidson 1983b; Stanley, Smith, & Howell, 1983). Thus, the bulk of the evidence is against an eye movement abnormality underlying dyslexia. The conclusion has to be that visual-spatial abnormalities are not a general cause of reading disability, although the possibility remains that there is a very small subgroup of dyslexics who have such abnormalities.

Dyslexia Subtypes

Most research and theory in dyslexia (and learning disabilities in general) emphasize multiple sources of disability. For example, it is widely believed that some children have a *perceptual* disability while others have an *auditory* disability (Johnson & Myklebust, 1967). Another influential view is that dyslexics have fundamental sensory disorders, either basically visual or resulting from failures at the "intersensory" level, i.e., visual-to-auditory connections (Birch, 1962). A system that includes three subtypes, but with visual disorders the *least* important, was proposed by Mattis, French, and Rapin (1975). These multifactor approaches, although they differ somewhat in what they identify as basic subtypes, have the obvious appeal of making individual differences rather basic. Their insistence that there is no single "cause" of dyslexia also has a strong commonsense appeal. Nevertheless, it is far from clear that any specific proposal concerning multiple causes of dyslexia has been able to obtain convincing research support.[2]

One intriguing proposal, in many ways typical of the subtype approach, is that of Boder (1971, 1973). Boder's proposal is in the spirit of others that assume at least two varieties of dyslexia. A very similar scheme was proposed by Ingram, Mason, and Blackburn (1970), but Boder's classification has attracted more attention. Boder (1971) described three distinctive patterns of reading disability: the dysphonetic, the dyseidetic, and the combined dyseidetic and dysphonetic. The origins of these terms, which were coined by Boder, suggest their meaning: *Dysphonetics* have a disability in phonetic processes. *Dyseidetics* (from the Greek *eidos*, shape or form) have a disability in holistic visual processes. The holistic aspect is important because Boder emphasized that problems in recognition of visual gestalts—whole words—are the basic characteristic of the dyseidetic. (This claim had previously been made by Bender, 1956, to apply more broadly to dyslexia generally.)

Dysphonetics, the most common dyslexic subtype according to Boder, read words based on "whole word visual gestalts." Their disability is evident when they encounter words not in their sight vocabulary, because they lack word analysis skills. Their decoding skills and linguistic skills in general are weak, and they read words better in context than in isolation. This description is reminiscent of the low-ability reader described in Chapters 5 and 6. The dysphonetic is also a poor speller, dependent on whole-word images. The combined dysphonetic and dyseidetic has both of these disabilities, experiencing problems both with holistic word perception and with phonetically based analysis. This combination subtype is the most handicapped, being unable to read "either by sight or by ear" (Boder, 1971, p. 314). In terms of the frequency of the three groups of dyslexia, Boder's observations were that dysphonetics were the most numerous, constituting about two thirds of the cases from a sample of 107 and being six to seven times more common than dyseidetics. A diagnostic screening test that distinguishes these two subtypes was published by Boder and Jarrico (1982).

This classification scheme for dyslexia has one especially interesting feature. Its largest category includes the features of ordinary low reading ability, that is, difficulties with the kinds of linguistic analyses that are needed for word decoding. This raises the possibility that there may not be a sharp boundary between dyslexia and ordinary low reading ability at the functional level. Any difference between them may lie elsewhere than in the cognitive and linguistic patterns of ability. Whether there are fundamental etiological differences (e.g., in the neurological bases for the disability) would remain an open question on this assumption.

The possibility suggested here is that there is a continuum of reading ability, with distinctions between dyslexia and low reading ability being largely quantitative. This suggestion runs counter to the prevailing opinion among disability specialists, who prefer a sharp distinction, typically with the assumption that dyslexia is a neurological anomaly, often with genetic components (Boder, 1971; Critchley, 1964). However, at the functional and descriptive level, as opposed to the etiological level, the quantitative difference model seems plausible. The only obstacle to this continuum assumption is the putative existence of a group of dyseidet-ics. There is little evidence in research on reading ability that there are low-ability readers who have this dyseidetic symptom. Either they are so rare that they are not uncovered in normal research populations, or they are indeed a qualitatively different group that appear only in true specific-disability populations. In either case, it is possible that the "dyseidetic" population is either negligible in size or that it will disappear under more rigorous investigations than have been carried out so far.

One study that suggests that the dyseidetic category may not be useful was carried out by van den Bos (1982). According to Boder (1971) dysei-detic children should be normal in auditory processes (or "analytic" processes; see next section). If so, dyseidetics should perform better than dysphonetics on a task of auditory memory for letters. In van den Bos's study, 52 dyslexic Dutch children, 9 to 10 years old, were classified ac-cording to the subtype categories of Boder's test. This test (Boder & Jar-rico, 1982) requires both reading and spelling by the child; the subgroup classification depends on a number of indicators, especially whether the children are able to spell unfamiliar words phonetically (at least in part). Those who can are classed as dyseidectics; those who cannot, as dys-phonetics. According to this test, 22 children could be decisively cate-gorized as dysphonetic, 10 as dyseidetic, and 17 as combined or mixed.

Subjects were presented with four letters either visually or aurally and asked to recall the list. The three dyslexic subgroups all performed more poorly than the control subjects on both the visual presentation and the auditory presentation. However, among the three subgroups there were no differences for either task. Dyseidetics did not perform better on au-ditory presentation, and dysphonetics did not perform better on visual presentation. Essentially the same kind of result was found in a memory scan test that required subjects to produce just one letter from a string of four submitted to memory. Dyslexic children as an undifferentiated group performed more poorly, especially when the letter probed was not the last one presented. In all cases, there was no difference among the subgroups for either visual or auditory input.

In a second experiment, these subjects made same-different judgments for letter shapes and letter names. The key comparison was between physical identity *(bb)* and name identity *(bB)* under instructions to re-spond *yes* if the letters had the same name. The difference between these two conditions is taken to be "name retrieval speed" and is associated with verbal ability (see Chapter 8). If dyseidetics are good at phonetic processes but not holistic visual ones, then their naming speed (*Bb* compared with *bb*) should be fast relative to dysphonetics. The latter group should be fast at the perceptual level *(bb)* but not on the name level *(Bb)*. However, the results were that the subgroups were *not* differentiated by either the perceptual level *(bb)* or the name level *(bB)*. All dyslexic groups performed more poorly at the name level compared with normals. They were actually slower than normals on all letter comparison tasks, but the name-same condition *(Bb)* indicated that dyslexics, as a group, had trouble especially with the name-level code. A similar conclusion was made by Ellis and Miles (1978) from experiments using this task. The important point again is a failure to distinguish dyseidetics from dysphonetics in a situation in which their putative specific disabilities might be expected to produce differences.

Of course, the fact that dyslexic subgroups did not perform differently on some particular experimental task does not mean that the subgroup classification is incorrect. For example, one might argue that performance on holistic word patterns, not individual letters, is a more appropriate test of the deficiency of the dyseidetic (and the strength of the dysphonetic). However, there is the possibility that the real difference between dysphonetics and dyseidetics is that they are two groups of disabled readers that differ in their spelling ability. It is the pattern of spelling errors that is a key indicator in the subclassification. It is possible that the subtypes of dyslexia identified by the labels "dysphonetic" and "dyseidetic" are reliable and even fundamental, but it remains to be proved.[3]

Auditory-Visual and Left-Brain–Right-Brain

The subtypes of dyslexia under discussion have usually been considered in terms of visual processes vs. auditory processes. Thus one subtype of dyslexia is based on an auditory deficit (dysphonetic), another subtype is based on a visual deficit (dyseidetic). Indeed, this dichotomy of visual and auditory subtypes is so prevalent as to be a commonplace assumption among disability specialists. Moreover, the kind of remedial treatment a child receives is supposed to depend upon which subtypes the child belongs to. Children with auditory deficits are to be taught with visually based instruction, whereas children with visual deficits are to receive auditory instruction (e.g., Boder, 1971; Johnson & Myklebust, 1967). But the idea that some individuals are visual processors whereas others are auditory processors does not seem very credible, or at least not very meaningful. Vision and hearing are, after all, simply fundamental sensory modalities for which normal functioning is assumed before the dyslexia label can apply.

More cerebral ways of looking at the dichotomy are *linguistic* vs. *non-*

linguistic and *analytic* vs. *holistic*. Equivalently, "left-brained" vs. "right-brained" refers to the same distinction. By this dichotomy, any modality (auditory vs. visual) differences among individuals is an incidental correlate of cerebral organization factors. In fact, an auditory-deficit individual should not have problems with all auditory inputs, only those that are linguistic. And a visual-deficit individual should have problems only with holistic spatial processing, while *word* perception, which is partly linguistic, should be normal. In fact, this scheme for a two-subtype dyslexia is gradually replacing the auditory-visual scheme, and Boder's dysphonetics and dyseidetics are consistent with this cerebral organization scheme.

The basis for a left-brain vs. right-brain classification is that the one cerebral hemisphere is specialized for linguistic and analytic processes whereas the other hemisphere is specialized for nonlinguistic and holistic processes. For most people, these are the left and right hemispheres, respectively. In addition to hemispheric specialization, one hemisphere is typically dominant, with the dominant hemisphere being the linguistic-analytic one. The evidence for specialization is quite strong, based on human subjects with severed interhemispheric connections (Sperry, 1964; Gazzaniga, 1970). In addition, differences in evoked potentials—measures of electrical activity on the scalp—show evidence for specialization, such as left-hemisphere activity for language tasks (Wood, Goff, & Day, 1971). There is also a vast body of evidence based on strictly behavioral measures such as dichotic listening—the right-ear advantage for speech—and hemifield visual perception—a right-hemifield advantage for words. (See, for example, Krashen, 1976, or Kinsbourne, 1978.)

Any subtyping of dyslexia based on left-brain vs. right-brain processing must assume differences in cerebral organzation, and of course this is essentially what Orton (1937) suggested. There have been many attempts since Orton to demonstrate that developmental dyslexia is associated with incomplete left-hemisphere specialization for language (Satz, 1976). However, the subtype approach has to demonstrate that one dyslexia subtype is associated with left-brain function—linguistic and analytic—and another with right-brain function—spatial and holistic (e.g., Bakker, 1979). Evidence is scarce, but at least one study reports that one group of dyslexics shows evoked potential patterns consistent with the Boder subtypes (Fried, Tanguay, Boder, Doubleday, & Greensite, 1981).

A variation on these dyslexic subtypes was proposed by Bakker (1979), who referred to Type I and Type II dyslexics. Based on dichotic listening and visual hemifield experiments, Bakker concluded that the Type I dyslexics showed typical left-hemispheric control of speech but ambivalent control of form perception by both hemispheres (atypical). Type II dyslexics are said to show right-hemispheric control of speech (atypical) and right-hemispheric control of form perception (typical). Thus each sub-type has an atypical dominance pattern, with one type having right-brained reading and the other left-brained reading. The error patterns made in oral word reading are claimed to reflect these dominance patterns, with the Type I making many form-based perceptual errors and the Type II making more linguistically based errors that preserve some syntactic characteristics of the target.[4] However, the correctness of these hemi-spheric subtypes, and especially their corresponding functional reading problems, remains to be firmly established.

Finally, there is another perspective on dyslexic subtypes. It is possi-ble, according to some, that the occurrence of subtypes is dependent on a child's development. Bakker (1979) suggests that children learning to read need right-hemispheric processes because of the demands of unfa-miliar letter forms. As children develop, the demands of reading become increasingly linguistic, and the language hemisphere takes on more im-portance. Thus, linguistic vs. visual dyslexic subtypes (or analytic vs. holistic subtypes) might be expected to emerge at different stages in learning to read. A similar proposal was made by Fletcher and Satz (Fletcher & Satz, 1980; Fletcher, Satz, & Scholes, 1981). They proposed that visual and spatial features dominate early reading and are respon-sible for differences in ability, whereas conceptual and linguistic factors dominate later reading and are responsible for ability differences beyond the early grades. However, they do recognize that linguistic factors are important even for early reading development, so the question is not as simple as it might seem: Are the linguistic abilities that contribute to reading ability at age 6 the same as those at age 11 (Fletcher, Satz, & Scholes, 1981)? It is difficult to obtain convincing evidence for such hy-potheses because of measurement questions and issues of theoretical analysis. For example, what counts as a linguistically critical skill at age 6 compared with age 11? And what aspect of reading is to be measured at these ages? Nevertheless, it is intersting that Fletcher et al. found that a measure of verbal fluency (the number of words the subject could pro-duce from a given semantic category within a time limit) was a better discriminator of dyslexics and normals for older children than for 6-year-olds. The use of English morphology (plurals, possessives, and other grammatical morphemes) distinguished normals from dyslexics at all ages. It is likely that different components of reading limit performance at dif-ferent stages of development, as we observed in Chapter 5. But there is no reason to doubt that some linguistic processes make an important contribution at all ages.

Right-brain dyslexia
So far, we have discussed the possibility of dyslexia subtypes based on

two distinct patterns of disability that are linked to patterns of hemispheric specialization. Thus, some dyslexics are claimed to be "right-brained" (dysphonetic) and some are claimed to be "left-brained" (dyseidetic). However, what we see is usually a highly disproportionate representation of the right-brained subtype. (This claim applies to Bakker's research only with certain qualifications; see note 4.) This disproportion suggests the possibility that most dyslexics are right brained, with a characteristic disability in linguistic and analytic processing. A truly left-brained dyslexic—one with genuinely normal linguistic and analytic processing—may prove to be a rarity.

A project reported by Gordon (1980) adds to this impression. Gordon studied 108 subjects, including 104 children between 8 and 15 years, who were referred to a reading disability clinic in Haifa, Israel. They were classic dyslexics in terms of the usual standards: average or above in IQ and 2.8 years below grade level in reading. (They were also 80% males.) Subjects were tested on a series of tests, some intended to tap processes of the left hemisphere and some intended to tap processes of the right hemisphere. The left-hemisphere tests included the following: (1) Serial sounds. Sequences from two to seven familiar sounds (e.g., a baby crying) were presented to a subject, who indicated the correct sequence of sounds. (2) Digit span. (3) Serial circles. The subject indicated the order in which sequences of three to five circles were "colored in" on a movie screen. (4) Word production. Subjects had one minute to produce (written or oral) words beginning with a given Hebrew consonant. The right-hemisphere tasks included these: (5) Form completion. Subjects named the common object presented as an incomplete silhouette, black on white paper. (6) Block design, the WISC intelligence subtest that requires the subject to reproduce a pattern of blocks. (7) Orientation. The subject had to decide whether two plastic models were identical when they were rotated in space. Notice that one of the left-hemisphere tasks had visual input and one of the right-hemisphere tasks required naming.

The most important result was that 105 out of the 108 dyslexic subjects performed better on the right-hemisphere tasks than on the left-hemisphere tasks. In a normal population, Gordon found that the average difference between right- and left-hemisphere tasks was zero.[5] It is not just that dyslexics were lower on the left-hemisphere tasks; they were actually higher than normals on the right-hemisphere tasks. In Gordon's interpretation, poor performance on the left-hemisphere tasks by itself was not related to dyslexia unless it was accompanied by better than average performance on right-hemisphere tasks.

The suggestion that dyslexics have such a cognitive asymmetry has been made before, based on low scores on verbal and sequential tasks and higher than average performance on so-called performance IQ items. Thus, dyslexic children are often above average on block design and below average on digit span (Naidoo, 1972; van den Bos, 1980). Explanations of how a lower verbal or sequential ability may contribute to reading failure are easy enough to construct. However, the significance of a "compensatory" skill in nonverbal or nonsequential tasks will have to be explained eventually if this right-brain profile turns out to be reliable across different populations.[6]

Summary

Among the many approaches to dyslexia, the assumption of anomalies in neurological development has been central. Virtually every functional and behavioral observation has been related to a hypothesis about cerebral functioning. The visual-spatial explanation of dyslexia seems inadequate as a general explanation of dyslexia, given the evidence. Somewhat more tenable is the assumption that there are dyslexic subtypes, one of which may include visual-spatial deficits. But the evidence for a specific visual-spatial deficit in the *absence* of a linguistic deficit is not compelling. It is clear that difficulties of left-hemisphere functioning (linguistic or analytic functioning) characterize the majority of dyslexics even when subtypes are identified.

VERBAL PROCESSES IN DYSLEXIA

It is clear from the various approaches to dyslexia that, whatever other subtypes appear, there is a clear association between reading disability and certain verbal processes. These may turn out to be rather the same ones as those that contribute to reading ability generally. The characteristics of low-ability readers prominently include difficulties with decoding and lexical access and working-memory limitations. The latter seems to be a problem not dependent on print, but a general processing problem. Do similar problems characterize dyslexic children?

Lexical Access and Decoding

Since the criteria for dyslexia include not being able to read words, there is no question about the existence of lexical access difficulty. The problem is how to explain it.

One approach to lexical access is to determine a child's word-recognition ability under brief presentations. Bouma and Legein (1977) tested 20 dyslexic Dutch children on recognition of tachistoscopically presented letters and words. The children were between 10 and 14 years of age and averaged three years below their age-appropriate reading level. The subjects viewed an isolated letter, a letter embedded between two x's, or a word of three to five letters. The stimulus was presented for 100 milliseconds in each case, sometimes foveally and sometimes parafoveally (1° either side of the fixation). The important results included these: The dyslexics had no difficulty with isolated letter perception. Performance was nearly perfect and there were very few of the classic dyslexic letter reversals. When the letter appeared parafoveally between two x's, accuracy was reduced and there were differences between dyslexics and normals. However, the dyslexics made errors on the same letters as did the normals. There was no evidence of an idiosyncratic perceptual problem. For word perception, dyslexics were less accurate than normals, but there were interesting additional facts. First, dyslexics made the same kinds of errors on the same words as the normals did. But dyslexics were more affected by word length, perceiving three-letter words accurately but making errors on four- and five-letter words presented foveally.

The conclusions from this pattern of results point to strictly quantitative differences between dyslexics and normals in word and letter perception. Dyslexics perceive isolated letters normally but begin to produce errors on letter strings and words, making errors similar to those that normal readers make, but more frequently.

In a later study with this population of dyslexics, Bouma and Legein (1980) measured subjects' vocalization (naming) latency to these stimuli. The results for correctly identified stimuli provided an important fact: Dyslexics were about 120 milliseconds slower at naming isolated and embedded (xax) letters than normals, but they were about 220 milliseconds slower for words. Thus, as is the case for low-ability readers generally, dyslexics show a longer decoding time, even when accuracy is controlled. Importantly, this decoding-speed factor seems to be something beyond a speed of letter processing factor. Bouma and Legein suggest that dyslexics have no strictly visual processing problems but have specific difficulty in the translation of print to a speech code.

This print-to-speech problem shows up clearly in a comprehensive study of dyslexics from age 7 to 17 by Olson, Kliegl, Davidson, & Foltz (1984). In one of several tasks performed by dyslexics and normals, subjects had to decide which of two printed nonwords sounds like a real word, e.g., *caik* and *dake*. Performance on this task was especially predictive of reading disability, even for older dyslexic subjects. There are many more studies that demonstrate some deficit in either speed or accuracy of word decoding and syllable decoding (see Vellutino, 1979). They largely confirm the picture of the dyslexic as not having rapid decoding skills at both the word and syllable level. However, a question that arises is whether there is something special about alphabetic inputs in the difficulties of the dyslexic.

The "naming" deficit

Denckla and Rudel (1976a) tested a large sample of dyslexic children ranging in age from 7 to 12 on tests of rapid naming. Children named colors, numbers, pictures, and words that were displayed on charts. On all tasks, dyslexic children were slower than normals and slower than learning-disabled children who were nondyslexic. Thus, this test of "rapid automatized naming" (RAN) seems to distinguish normals and dyslexics on general name retrieval. The speed factor is not restricted to printed words. This is not always the case for low-ability readers, as we have seen in the study of Perfetti et al. (1978) in which low-ability readers were slower only in words (and digits slightly) but not colors.[7]

It is likely that Denckla and Rudel have identified a dyslexia that does indeed reflect a generalized alexia of the sort seen in patients who have suffered left-hemisphere lesions. "Finding" names and producing them is difficult. Support for this interpretation comes from another study by Denckla and Rudel (1976b) which examined naming responses to line drawings of objects. Not only did dyslexics make more errors than normals and nondyslexic learning-disabled subjects, they made more errors of circumlocution rather than errors of misnaming or omission. They often knew something about the object to be named but could only describe it in terms other than its name. This is consistent with a pattern of some neurological dysfunction, and is reminiscent of aphasic disorders, as Denckla and Rudel observe.

The particular components of rapid naming that are responsible for the dyslexic's naming deficiency, however, are not clear. There are several components involved in rapidly naming objects displayed on a chart. Name retrieval is one, but there is also visual scanning and input-output sequencing. The role of sequencing may be significant in task performance and may be a source of difficulty for dyslexics. The subject must not only retrieve a name of a visual stimulus but also produce it while preparing to encode and name the next stimulus. In other words the critical task demand as far as the dyslexic is concerned may be the *rapid* sequential naming, not the *speed* of name retrieval. In any case the rapid-naming test is clearly a powerful test of dyslexia since it can distinguish dyslexic children from other disabled children as well as from normals.

Memory Processes

Low-ability readers, both children and adults, are distinguished by a reduced functional working-memory capacity. The qualification "functional" is to suggest that the mere passive memory storage is not a distinguishing factor, but an active processor is (see Chapter 5). In the research on dyslexia, there is also ample evidence that dyslexic children have reduced memory spans (Farnham-Diggory & Gregg, 1975; Guthrie, Goldberg, & Finucci, 1972; Moore, Kagan, Sahl, & Grant, 1982; Torgesen & Goldman, 1977). (Some studies, however, show no difference in memory span [Valtin, 1973]). The usual way to measure span is with digits, although letters, words, and pictures also produce differences.

A study by Moore, Kagan, Sahl, and Grant (1982) typifies the kinds of results found in many studies (see Torgesen, 1978, for a review). They had a stringently defined sample of dyslexic subjects and a carefully matched normal control group. A series of memory span tests required subjects to recall orally presented words, written words, and pictures. The span measure (the longest string recalled without error) distinguished dyslexics from normal subjects for all three categories. This result held true both for free recall (order irrelevant) and ordered recall instructions. In another task dyslexics were found to have digit spans lower than those of the normal readers. Finally, dyslexics also showed a memory deficit in a digit-recognition task in which a same-different judgment was made for two successively occurring digit sequences.

Thus, although it is true that poor memory performance does not always distinguish dyslexics, the bulk of the evidence suggests a general memory deficit. The memory deficit seems to be most observable when active processing is important. Recall that low-ability readers studied by Perfetti and Goldman (1976) did not show a deficit in a probe digit task (Chapter 5). This task requires a more passive recording of information, and active rehearsal is difficult to perform. The possible importance of active processing is demonstrated in the Moore et al. (1982) study, described above. Dyslexic subjects were trained in rehearsing digits while they were being presented; following this very brief training, they showed dramatic gains in digit memory performance.

A different explanation for why memory deficits are sometimes not found in dyslexics is that only a subgroup of dyslexics has such a problem. Torgesen (1982) reports finding one group of children with subnormal digit span and another with normal digit span, despite their being comparable in other respects. However, it appears that only the group with low digit spans really behaves like a dyslexic group on most tasks. Overall, the conclusion seems to be that short-term memory limitations characterize dyslexia and low-ability reading generally.[8]

The ordering function of memory is sometimes singled out for special importance in the dyslexic's memory deficit. Dyslexics perform more poorly with respect to temporal-order memory on all the kinds of materials mentioned above and also on nonspeech rhythm patterns presented visually or aurally (Zurif & Carson, 1970). Recall also that one of the dyslexic diagnostic tests of the Gordon battery was temporal ordering of visual patterns. However, although it is conceivable that a special temporal-ordering deficit exists independently of memory capacity, it seems reasonable to speak of a single functional memory problem. Ordering of information happens to be a prominent feature of memory during processing of spoken and written language. Accordingly, it is a large component of the reading task that makes demands on the dyslexic's limited functional memory capacity.

Linguistic and Speech Factors

The verbal processes of reading include some linguistic and speech factors that can be involved both in the decoding and memory deficits of dyslexia. The main factor has to do with the use of speech codes in decoding and memory. We will focus on speech processes after a brief consideration of other linguistic factors, specifically syntax.

Syntactic abilities in dyslexia

There is no doubt that dyslexic children fail to acquire the syntactic abilities of their peers. It would be surprising if it were otherwise. The question is whether a syntactic disability is basic or derivative. It is likely that syntactic abilities beyond those the child already possesses at the age of 5 or 6 are considerably modified by reading and academic learning.

Dyslexic children, however, show problems even on simple syntactic structures that should be acquired early. Young dyslexics, ages 5–7, tend to show problems with the English inflection system, and so do even older dyslexics of age 11 (Fletcher et al., 1981). Vogel (1975) found that 8-year-old dyslexics also had problems with simple inflections such as noun pluralization, ordinarily acquired in the first three or four years. Given that simple morphological structures are a problem for both younger and older dyslexics, it should not be surprising that both have problems with more sophisticated syntactic structures or that their speech shows a less mature syntactic structure (Fry, Johnson, & Muehl, 1970).

It is possible that linguistic knowledge is fundamental in providing cognitive structures to which sentence meanings can be attached. However, the dyslexic has a more basic problem: the inability to read words. If syntactic abilities are critical to word reading, it may be because there are analogous linguistic processes involved. One possible link between syntax and word reading is that both involve a linguistic rule system. (In fact, Morrison, 1980, suggested that dyslexia can be viewed as a general rule-learning deficit.) In any case, it is not necessary to explain dyslexia by reference to a syntactic deficit. It is important, however, to emphasize the general coincidence of linguistic disabilities.

Speech processes

Reading involves speech processes (Chapter 4). Thus it is possible that the use of speech codes in reading may be a problem for dyslexics. The most prominent possibility is that beginning readers need to use speech codes to learn to read and to support their early reading. We may well expect young dyslexics to be children who fail specifically at this part of reading, which is critical for decoding. However, we will postpone consideration of this until the next chapter, where we consider the beginning reader. For now, we ask whether the disabled reader beyond the beginning level has a speech-based deficit. To do this we refer to two basic possibilities, nonactivation and deactivation.

The possibilities for speech deficits among dyslexics are clearly implied by the identification of mild aphasic symptoms in some cases (Denckla & Rudel, 1976a). Also, they are implied for any subgroup identified as linguistically impaired, such as Boder's dysphonetics. But determining exactly what kinds of speech deficits might exist, especially for a relatively unselected dyslexic population, is difficult. Such a deficit, by definition, would have to be more than mere auditory sensitivity. It could involve anomalous *speech* perception (not mere auditory perception), difficulties in speech-print mappings, deficits in speech memory, or all three. In terms we have used within the activation model (Chapters 4, 5, and 6), the dyslexic reader may suffer from nonactivation of speech codes from print or the deactivation of speech codes in memory. Or he may suffer from a general speech perception problem.

Memory deactivation
There is evidence that young disabled readers (second grade) have speech memory functions different from normal. The suggestion that phonetic memory codes are not used by young disabled readers was made by Liberman, Shankweiler, Liberman, Fowler, and Fischer (1977) and received support in a number of experiments. Liberman et al. (1977) visually presented second-grade subjects with strings of five letters for three seconds. Subjects reproduced the letter strings following 0- and 15-second delays. Some letter strings rhymed *(B, C, D, G, P)* and some strings did not *(H, K, L, Q, R)*. The basic result was that rhyme letters were recalled less well than control (nonrhyming) letters, especially when there was a 15-second memory interval rather than immediate recall. However, the key result for reading ability was that the rhyming effect was smaller for low-ability readers than for skilled readers. That is, skilled readers did much more poorly when the names rhymed than when they did not, but low-ability readers were not much affected by rhymes. Liberman et al. (1977) concluded that low-ability readers showed fewer confusions due to rhyming because they were not as good at phonetic coding.

Since the original results of Liberman et al. (1977), this ability difference in the rhyming effect has been extended in three directions: (1) It has been found in listening as well as reading (Byrne & Shea, 1979;

Shankweiler, Liberman, Mark, Fowler, & Fischer, 1979); (2) It has been found for words and sentences as well as letters (Mann, Liberman, & Shankweiler, 1980); (3) It has been found in recognition memory as well as recall (Byrne & Shea, 1979; Mark, Shankweiler, Liberman, & Fowler, 1977).

It is especially significant that the rhyming effect holds for speech as well as print. It strongly suggests either a basic process factor in phonetic coding or a strategy factor for dealing with language tasks. In either case, it does not require an explanation particular to print decoding.

It is not clear, however, how general the relationship is between reading ability and the rhyme effect. Olson, Davidson, Kliegl, and Davies (1984) find that this relationship holds for second-grade disabled readers but not for older readers. In research with the Pittsburgh low-ability sample, we find no relationship between phonetic memory and ability beyond the second grade.[9] This is not likely to be a difference in level of reading disability. The samples of Olson et al. seem to fit the usual definition of dyslexia, and the samples of the Pittsburgh studies appear comparable to the Haskins studies (Liberman et al., 1977) and those that followed. Apparently the use of phonetic codes in memory is a more distinguishing factor for young dyslexics than older ones, at least as far as the rhyming tasks measures the use of phonetic codes. In fact, the reduced use of name codes by the low-ability reader can be seen in other tasks. For example, Katz, Shankweiler, and Liberman (1981) found that memory differences between high- and low-ability readers were larger for nameable drawings than unnameable drawings.

Speech perception
That dyslexics might be less likely to use speech codes in memory can be understood in a number of ways. They are less good at decoding, less sensitive to the structure of language, and perhaps less likely to use a verbal rehearsal mechanism. However, it is possible that dyslexics have fundamental speech perception anomalies. At least this is the rather surprising conclusion from a study by Godfrey, Syrdal-Lasky, Millay, and Knox (1981). They tested severely dyslexic children of age 10 on the identification and discrimination of synthesized speech syllables. These were stimuli from a continuum from /ba/ to /da/ and from a continuum of /da/ to /ga/. Successive stimuli from each continuum differed by a constant change in starting frequencies of the synthesized formats.[10] The key observation concerned phoneme boundaries. When normal subjects hear such stimuli and are asked to label each syllable (as, say, "ba" or "da"), there is typically a sharp category boundary. For example, the three stimuli closest to the end point /ba/ are called "ba," and the three closest to /da/ are called "da" with very little mixed responding. Surprisingly, Godfrey

et al. found dyslexic subjects not to have the normally sharp boundary, at least for the /da/–/ga/ continuum. In other words, their syllable categories were less "all or none." Moreover, their discrimination functions were unusual. Normal subjects tend to show categorical perception. That is, they are more likely to perceive two stimuli as "the same" if they are within the same phoneme category than if they are from different phoneme categories, even though all pairs to be judged are equally different in terms of acoustic parameters. Dyslexics were less likely to show this categorical perception pattern, at least for the /da/–/ga/ continuum. Generally, they were poorer at detecting a difference between two stimuli even when they were across a phoneme boundary. Thus, dyslexics appear to be less consistent in their identification of at least some syllables and less able to detect differences between slightly different examples of these same syllables.

What all this means is difficult to say. If the results are reliable, they point either to a speech-processing mechanism that is distinctly nonnormal at the phoneme level or to some spurious task-specific problem. It is important to emphasize that these are not general auditory differences but processes relating to acoustic properties necessary and sufficient for speech perception.[11] Of course, this does not mean that some other "auditory" factor cannot be found distinguishing dyslexics from normals. But it is a good bet that only acoustic factors that have an important role in consonant perception will be significant—for example, acoustic parameters relating to rapid changes in frequency. It would be premature, however, to conclude that there are fundamental differences in speech perception mechanisms between dyslexics and normals.

Finally, whatever the interpretation of the speech perception question, it is interesting that in the Godfrey et al. (1982) study, the investigators were able to select two groups of dyslexics by using the Boder test. The dyseidetics and the dysphonetics did not differ in these phonetic tests, a finding that adds to the weight of negative results that cast doubt on the existence of two fundamentally different dyslexic types. It may well turn out that dyseidetics and dysphonetics are not fundamentally different but merely represent two adaptive and strategic variations of a single dyslexia.

Dyslexia and Reading Ability

One of the questions raised in this chapter is whether there are factors in dyslexia comparable to those that are important in low-ability generally. With the possible exception of the research of the Haskins group (Liberman et al., 1977) on speech factors in memory, the research referred to in this chapter fits a fairly stringent definition of dyslexia. Careful definitions are especially important where neurological signs are concerned, and the severity of the disability is an important factor. Allowing for these considerations it seems that factors contributing to dyslexia are much the same as those that contribute to less specific examples of low reading ability. We may say that within the limits of general intellectual ability, reading ability is a continuum. A child who at age 10 is three years below his or her expected grade level is more disabled than one who is one and a half years below. These differences in disability seem to be more of degree rather than kind. Both groups will have trouble with decoding, and both have reduced working-memory capacity. Both will also show low skill in general linguistic processing ranging from syntax to phonetic processes. The one area that provides plenty of room for disagreement is that of phonetic processes, specifically, the origin of observed phonetic differences between low-ability readers (and dyslexics) and normals.

There is a problem with this attractive simplification, however. If there are dyslexic subtypes, then dyslexia itself cannot represent a point on an ability continuum. We have referred to the two subtypes that are most prominent in dyslexia, the dysphonetic and the dyseidetic. It would be premature to say that the two subtypes do not exist. Their existence has some plausibility because they are based on generally differentiated processes, each of which could be associated with consistent individual differences. If evidence becomes stronger for these two subtypes as fundamental, then our reading ability continuum will have a discontinuity. There will be a continuum of linguistically based ability that includes the dysphonetic at one end and the low-ability readers throughout the low end. (It may even include the adults identified by J. Baron, 1977, as "Chinese"—readers dependent on holistic rather than orthographic word recognition.) The dyseidetic would remain an example of discontinuous specialized form of dyslexia, rather a curiosity from the standpoint of general ability. In any case, there is sufficient doubt about the dyseidetic to keep open the possibility that there is a single continuum of ability.

It is possible that some of the attraction of dyslexic subtypes comes from acquired dyslexia. Case studies of patients who have suffered strokes or brain lesions describe some fascinating symptoms. Some cases suggest loss of a phonetic ability and some suggest loss of a semantic ability, leading to the conclusion that semantic and phonetic abilities are completely independent of each other (e.g., Marshall & Newcombe, 1973; Shallice & Warrington, 1975). That is, some patients seem to have meanings activated by words without the word's phonetic properties (deep dyslexia), while others can decode but without meaning (surface dyslexia). In fact, Coltheart (1980) suggests some eight specific subtypes of acquired dyslexia. Certainly, the identification of functional independence of phonology and semantics, for example, is important for neuro-linguistic models of reading and language. However, it is not clear that acquired dyslexia should provide a model for developmental dyslexia. Of course, developmental dyslexics may show neurological disturbances just as acquired dyslexics do. Geschwind (1979) reports a postmortem study (by Galabdura and Kemper) of an adolescent developmental dyslexic (who was also epileptic) that revealed a marked but highly selective structural abnormality in the cerebrum. It was restricted to a left-hemisphere structure mapped for language function, leaving the corresponding right-hemisphere structure unaffected. Such evidence suggests that developmental and acquired dyslexia can eventually be understood as sharing neurological causes. For now, however, it seems prudent to assume that developmental dyslexia needs to be understood on its own terms. Certainly its subtypes must be motivated on behavioral grounds.

Finally, on the question of subtypes, suggesting a continuum is not the same as claiming that there is a single cause of dyslexia. There can be many different combinations of problems, perhaps even enough to warrant subclassification for some descriptive purpose. However, the continuum idea would mean that a common thread runs through all the problems. That thread is the linguistic processes that are needed for decoding, verbal memory, and speech understanding.

SUMMARY

Dyslexia, or specific reading disability, is a developmental disorder with a presumed neurological basis. Although developmental dyslexia must be distinguished from acquired dyslexia, which is the result of cerebral insult, it may not be very different from low reading ability in general. In both cases, the reading problem is not a matter of general intelligence but of factors that, more or less specifically, affect reading and language.

Traditional views of dyslexia emphasized visual-spatial deficits. The bulk of evidence counts against this view, although dyslexics sometimes show problems that seem to be visual-spatial. Increasingly, dyslexia has been seen as a nonhomogeneous label with two, three, or four subtypes. One of the subtypes is usually visual-spatial and the other is auditory-linguistic. Boder's dyseidetics and dysphonetics are examples of such a subtyping. Evidence for two distinct subtypes is not yet compelling. Another view is that dyslexics are generally defective in linguistic-analytic ("left-brain") processing. The auditory modality is not essential.

Regardless of subtypes, a major fact of dyslexia is its verbal component. Problems of lexical access and decoding are severe and almost definitional. These problems have a component beyond simple letter processing, suggesting actual decoding problems similar to those of low-ability readers. Dyslexics also have naming problems independent of print, and memory problems. Syntactic factors play a role in dyslexia, although probably a secondary one. Speech processes may be more fundamental, at least for young children. However, the origin of speech problems, especially in speech perception, remains to be understood.

Finally, there seems to be reason to think that dyslexia may be the low end of an ability continuum rather than a qualitatively different problem. The severity of the disability is the main factor in whether one is dyslexic. The continuum hypothesis will gain support if the identified subtypes of dyslexia turn out not to be less fundamental than has previously been assumed.

NOTES

1. It is beyond our purpose to review and evaluate all the evidence claiming to show that visual and spatial deficits underly dyslexia. Vellutino (1980) provides a very thorough critical review of this literature.
2. Birch's theory has probably been the most successful in generating meaningful research. See, for example, Birch and Belmont (1965) and Belmont and Birch (1966).
3. There are studies which seem to support dyslexic subtypes of this sort. Aaron (1978) is often cited in this regard, although his study is not based on Boder's classification. A study by Fried, Tanguay, Boder, Doubleday, and Greensite (1981) reports that dyseidetic and dysphonetic subjects show different patterns of evoked potentials to musical and word stimuli. On the other hand, Davidson, Olson, and Kliegl (1982) in a study of 141 dyslexics from age 7 to 17 found no evidence for a dyseidetic subgroup.
4. The error patterns Bakker (1979) reports are rather complex and in fact it is not possible to examine the types of errors in detail nor to be sure that error rates are comparable across the two subtypes. However, the linguistic nature of the Type II error in oral reading is said to be reflected in "time-consuming errors"—corrections, repetitions, and corrections of misreading. Type I dyslexics, who seem to make fewer errors in oral reading, make relatively more "substantive" errors—omissions, replacements, inversions. It is not clear that the two types differed in the visual similarity between their errors and the target words. It also appears that Type I is much more frequent in Bakker's data. It is easy to believe that right-hemisphere language specialization (Type II) is relatively rare. It is somewhat more difficult to be convinced that linguistic errors are *less* common among dyslexics than "visual" errors, given other literature on the patterns of disabilities shown by dyslexics (e.g., van den Bos, 1982; Gordon, 1980). It seems appropriate to conclude that the Bakker subtypes should not be accepted without some reservations. Of course, the same is true for the Boder subtypes.
5. To allow these comparisons, each normal subject received a score on each test. Each test score was then normalized to a Z score, giving each sub-

ject an average Z score for left-hemisphere tests (P) and one for right-hemisphere tests (A). The difference between these two averaged Z scores (A − P) is what Gordon called the "cognitive laterality quotient." This difference score has a mean of 0 for the normal population and a mean of 1.12 for dyslexics. If A and P are added, rather than subtracted, the two groups are nearly identical.

6. One possible explanation for its significance may lie in strategy differences associated with reading ability. Caplan and Kinsbourne (1981) report that a preference for a verbal approach to problem solving is associated both with a child's reading ability and left-hemisphere processes. Conceivably, a right-brain profile may indicate a person who approaches problems, including reading, visually.

7. Spring and Capps (1974) found a large deficit for dyslexics in digit-naming speed and a smaller deficit in color-naming speed. Although we think there are subject population differences across various studies that help explain discrepant results, there is a critical methodological difference. In the Perfetti et al. (1978) studies, what is measured is the time to name an individually presented item—color, digit, word, or picture. In Denckla and Rudel and in Spring and Capps, subjects name sequences of items from a chart. This is a significant difference. The individual-item procedure measures naming time. The chart procedure measures sequencing and resistance to interference in addition to naming time. It may be significant that Denckla and Rudel as well as Spring and Capps both found naming colors to be slower than naming digits, whereas Perfetti et al. (1978) did not. The chart arrangement can easily produce a Stroop-like interference in which what the subject has just *said* is interfering with what he is looking at. (The Stroop interference effect refers to the difficulty in naming colors that are incompatible with the color names containing them, e.g., the word *red* printed in green ink.) It may be that rapid sequencing rather than rapid naming is the key to the naming deficit.

8. Richardson (1982) also reports finding two subgroups. One was low on digit span relative to controls and one was not. The one not low on digit span was also not low on verbal IQ and vocabulary measures. A complete disassociation of these two verbal abilities would be odd.

9. There have been three studies with the Pittsburgh population that find at most small differences between high- and low-ability readers in phonetic-based confusions in memory. In two experiments, no differences at any age were found in a recognition memory experiment with forced-choice responding, which eliminates response bias (Beverley & Perfetti, 1982, report one of these.) In a free-recall experiment, there was no difference between high- and low-ability readers in the rhyme effect except when an order-sensitive measure was employed (Leahy & Perfetti, 1984). In fact, high-ability subjects in the recall study recalled more rhymes then did low-ability subjects.

10. For the /ba/–/da/ continuum the starting frequencies of the second and third formants were increased in equal steps. For the /da/–/ga/ continuum, third-formant starting frequencies were decreased in equal steps. These formant changes are sufficient acoustic cues to place of articulation for the stop consonants, /b/, /d/, and /g/.

11. A study comparing third-grade high- and low-ability readers (not dyslexics) by Brady, Shankweiler, and Mann (1983) is interesting in this respect. They found that low-ability readers made more errors than high-ability readers when listening to speech but not when listening to environmental sounds under conditions of noise. This seems consistent with a perception problem unique to acoustic information in speech.

PART THREE

Beginning Reading and Reading Instruction

CHAPTER TEN

Beginning Reading

How does a child learn to read? That is the general question addressed in this chapter. Since our focus has been on ability differences, we also ask the corollary question: What ability differences emerge during beginning reading? Finally, because we have suggested that skill in reading is largely a matter of verbal efficiency, we ask whether verbal efficiency factors are observed in beginning reading.

LEARNING TO READ

There are many ways to begin reading. The way a child begins reading depends on at least two general factors: the writing system and the instructional methods. It also depends to a considerable extent on very individualized factors, especially preschool reading experiences at home.

Writing Systems and the Alphabetic Principle

Learning to read Arabic is different from learning to read German or English, and all three are different from learning to read Chinese. The critical difference among writing systems is how the units of the writing system, the script, code the units of the language. At one extreme is a writing system that directly codes meaning. Early picture writing offers the earliest examples of such a system, but more abstract representations, or ideographs, are more common in meaning-based writing systems. In fact the evolution of writing systems seems to have moved away from meaning-based systems and toward alphabetic systems (Gelb, 1952; Gleitman & Rozin, 1977). Chinese remains the most widely used meaning-based writing system in which ideographs and pictographs are prominent.

For a reader, the critical fact about an ideographic system is that associations must be learned between units of meaning—words and morphemes—and the units of writing. In the case of Chinese, the writing units are characters that include simple ideographs, pictographs, compound ideographs, and phonetic compounds, the largest category of Chinese characters.[1] While the different character types reflect different histories and constituent makeups, they share this important feature: Some part of the character corresponds directly to a word meaning in the language. Thus, the child learning to read must learn to associate characters to word meanings.

The degree of literacy in Chinese depends on the number of these associations that have been learned. A vocabulary of between 5,000 and 7,000 characters may characterize literate Chinese adults, although children in elementary schools are expected to master only about 3,500 characters by learning about 500 or 600 per year during each of six years in school (Leong, 1973). In contrast to their American counterparts, Chinese children spend significant time at home working on Chinese characters, perhaps a necessary condition for literacy.

The existence of ideographic writing systems, although they are rare, demonstrates that learning to read can take place by association of symbol with meaning. Equally clear, however, is the value of a writing system that requires fewer than 3,000 or 4,000 associations to be learned. The two systems that allow this economy of learning are syllabaries and alphabets. Each gains economy by creating correspondences between units of writing and units of speech. They take advantage of the fact that speech is already associated with meaning in the language.[2] A syllabary system associates written symbols with syllables, and an alphabetic system associates written symbols with abstract phonemes. Syllabaries, historically prior to alphabets, are relatively rare today, Japanese being one clear example. As Gelb (1952) put it, the alphabet, having evolved from syllabary systems, "conquered the world."

In fact, the various alphabets of the world, despite visual differences, all operate on exactly the same principle. The original Greek alphabet, the Roman, Cyrillic, and Arabic all express the *alphabetic principle:* A written symbol is associated with a phoneme. Some alphabets reflect their syllable origins more clearly than others by not having letters for vowels, for example, Arabic and Hebrew. But the principle of association is the same and this principle is the key to a *productive* writing system. A productive writing system is one that produces the indefinitely large number of words and morphemes from a small set of reusable symbols. The 26 letters of the Roman alphabet, the 24 letters of the Greek alphabet, and the 32 letters of the Russian alphabet all represent very productive writing systems.

Obstacles to using the alphabetic principle

The productive value of the alphabetic principle is its essential glory. However, it is apparently also a source of distinct discomfort. Discovering and using the alphabetic principle is not always easy for beginning readers. There are at least two obstacles: the abstract nature of the phoneme, especially consonants; and the failures of alphabets to code each of their vowels with a unique written symbol. Both of these factors can make learning to read a bit of a challenge.

The abstract nature of the phoneme is an intrinsic characteristic of speech (see Liberman, Cooper, Shankweiler, Studdert-Kennedy, 1967). It is true of all languages. However, it is a problem only for consonants because vowels, while abstract in a strict sense, are relatively "concrete." They last long enough to hear and they are less dependent on context. The /ae/ in *rat* is about the same as in *laugh* and in each word the vowel sound can be clearly heard and isolated. This means someone, for example a teacher, can point to the *a* in *rat* and say "aah" (/ae/) without being deceitful. There really is an /ae/ sound in *rat*, and a child can hear it because it lasts long enough and it does not depend much on what sounds precede or follow it.

This is not the case with consonants, especially stop consonants. The /d/ in *dime* is not quite the same acoustically as the /d/ in *dome* or the /d/ in *lid*. The phoneme /d/ is an abstraction. It exists only as a perceptual prototype of all the [d] sounds that actually occur. Furthermore, its perception and production are highly dependent on vowels that precede and follow it. In fact, when a teacher tries to say that the word *dime* begins with /d/, there will be a problem. She will produce /d/ plus a bit of a vowel because some vocalization is necessary. If she adds the vowel sound /ay/ ("eye"), thus producing /day/, that's correct for *dime* but not for *dome*. If she tried to omit the vowel, producing /d/ or /də/ ("duh") to correspond just with the letter *d*, that's incorrect for both *dime* and *dome*. However, it may be closer to the consonant prototype because it includes a minimum vocalization.

This means that "discovering" phonemes such as /t/, /d/, /p/, and /b/ is especially difficult. It follows that applying the alphabetic principle may also be difficult because it depends on associating the abstract /d/ with the letter *d*. Associating two elements will be difficult if the child has only one of them.

The second difficulty in applying the alphabetic principle is less fundamental. For alphabetic writing systems, the price of being economical is to have complex correspondences for vowels. American English has about a dozen vowels but only five standard vowel letters. That means that *A, E, I, O, U* have to do quite a bit of double and triple duty, even with the help from *Y*. For example, *bat, bar, bake,* and *fiat* each use the letter *A* for a different vowel phoneme. When an alphabet allows this to happen, it compensates to some extent by complicating the orthography. For example, the pronounciation of *a* in *bake* is determined by the presence of a final *e*. And the pronunciation of *a* in *bar* is determined by the presence of the *r* and an absence of a final *e*. Thus it is far from the truth to say, as some have, that English spelling is chaotic. The regularity of pronunciation, which is the key advantage of a productive alphabet, is largely present. However it is not always present in a one-to-one mapping of letters and phonemes.

The reason for saying that this vowel problem is not fundamental is that it is not intrinsic to language. It is a result of alterable orthographic principles and it varies from one orthography to another across languages. For example, the German alphabet is only trivially different from the Roman alphabet, except that it encodes fully the German vowels by use of diacritical marks (umlauts: *a* vs. *ä*). In principle, it is possible to do the same kind of thing for any language. Generally, there is a trade-off between the explicitness of the writing system, which is increased by more graphic symbols, and the economy of the writing system, which is decreased by more graphic symbols.

In addition to the economy vs. explicitness trade-off, there is also a trade-off of explicitness and morphological transparency, that is, between the explicitness of the writing system in representing the *phonemes* of the language and the transparency of the writing system in representing the *morphology* of the language. A fully explicit system, for example, could associate *a* only with /ae/ (as in *mat*) and invent some other symbol for /e/ (as in *mate*). However, this would mask important morphological facts that are reflected in the present system. For example, *nature* and *natural* use *a* to represent two different phonemes. But this *a* spelling reminds the reader of their noun-adjective relation and is typical of English morphological spelling.[3]

Since English orthography is one that sacrifices explicitness for both symbol economy and transparency, there have been proposals to alter it for the purpose of teaching reading. The Initial Teaching Alphabet (ITA) of Pitman is the best-known example. It has 44 written symbols, one for each English phoneme, including 16 vowels. In an effort to retain some economy, Pitman used a number of digraphic symbols. For example *oi* symbolizes the vowel in *boy* and *ie* symbolizes the vowel in *kite*. There are some 14 of these digraphic symbols, producing a bit of visual clutter as the price of explicitness. Of course, the point of the digraphs is that they retain one of the letters of the regular alphabet that the child will eventually have to use. There have been inconclusive studies of whether the ITA helps a child learn to read (Downing, 1964; Chall, 1967). However, any failures to teach beginning reading that might occur in the use of an explicit alphabet must be purely technical—for example, the intricate procedures of transfer to the normal alphabet (see Holland, 1979). Because such an alphabet explicitly codes all phonemes in a consistent manner, it has to enable easier applications of the alphabetic principle.

One other solution to the difficulty of learning to apply the alphabetic principle is the syllabary. A syllabary system has an intermediate speech level between the written symbol and the word level. However, it does not have the intrinsic problem of phoneme abstractness that the alphabetic system faces. Syllables have acoustic duration and they provide a more or less invariant unit. These facts concerning syllabaries, along with the fact that syllabaries remain as the basis for some writing systems (e.g., Japanese Hiragana), prompted an attempt to teach reading by syllabary undertaken by Gleitman and Rozin (1973). They reported success in teaching children with low reading prognosis by means of a rebus-based syllabary. A rebus is a picture that suggests its name. The rebus name, in a syllabary instruction, is then used as a syllable. For example, a picture of a can is first used for the word *can* and then for the syllable *can*, as in *candy*. The point of their demonstration was that early instruction in reading can introduce the child to the speech-based nature of reading. It is not a substitute for an alphabet; it is merely a means to help the child discover that the print encountered in reading corresponds to speech and not to meaning. It is important not to miss the fact that the alphabetic system and the syllabary system both encode speech. The alphabetic system is much more efficient (there are thousands of English syllables) and is the target writing system to be learned, regardless of initial teaching geared toward making the alphabetic principle easier to apply.

In summary, there are indeed some potential problems in discovering and applying the alphabetic principle. The intrinsic theoretical problem is that phonemes are perceptual abstractions. The extrinsic problem is that some alphabets may sacrifice phoneme explicitness for symbol economy or for morphological transparency, thus complicating the orthography. However the essential value of the alphabetic principle is as a productive symbol coding system. Once mastered, it allows the reader, even the young reader, to read (and write) words never seen before. Only a writing system that has a mediating speech level between written symbols and language meanings can do this.

Linguistic Knowledge and Learning to Read

The discovery and application of the alphabetic principle, as the preceding section suggests, is not easy but it is surely important. As Ehri put it, "If the light were not so gradual in dawning, the relationship between speech and print might count as one of the most remarkable discoveries of childhood" (Ehri, 1979, p. 63). The general cause of these lingering predawn mists has to do with "linguistic awareness," or rather the lack of it. One particular example of linguistic awareness has already been illustrated. The alphabetic principle depends on associating letters with phonemes, but the abstractness of the phonemes makes them difficult to notice. Thus, *phonemic awareness* is one particularly important kind of linguistic awareness.

In general, we mean by "linguistic awareness" knowledge of linguistic structure that is at least partly accessible. It is, in essence, the knowledge of linguistic form, as opposed to meaning. Every level of language has its own structure: phonological, morphological, and syntactic. Writing introduces another structure, namely orthography. In each case, some aspect of reading depends on knowledge of the linguistic structure, beyond the comprehension and production of messages. For example, to have morphological knowledge is to know that *ate* and *jumped* share a formal structure despite the absence of shared meaning and shared phonemes. It also implies the ability to see the structure in new forms or even in nonsense forms, e.g., *glumped*.

This kind of linguistic knowledge does not have to be a product of *awareness* at all. Rather it has to be implicitly available to a child in his or her encounters with everyday language. Hence *implicit knowledge* may be a more accurate characterization of what a child knows about language structure. However, learning to read requires some kinds of linguistic structure to be brought up to a partly explicit level of knowledge. If the teacher says the first sound of *apple* is "aaah" (/ae/), the child must mentally operate on something that corresponds to the word *apple* and something that corresponds to the phoneme /ae/. (The process would be more difficult with a consonant phoneme.) A child who has some knowledge that words are independent linguistic objects and that they consist of meaningless speech sounds should be a better bet for learning to read than a child who does not. In the following sections, we will examine the kinds of linguistic knowledge a child possesses at the start of reading instruction.

Knowledge of Speech Sounds

The communicative nature of language seems to conspire against form knowledge. Children, prior to beginning school, are reasonably competent users of language for communication. They can produce and comprehend speech in communication situations. However, the formal aspects of language, including the phonetic structure of speech, have had less chance to develop. The phonetic level is relatively "transparent"; that is, it is easy to not notice it. It is not surprising that knowledge of language function precedes knowledge of language form, but reading requires at least some knowledge of form. Phonology must become somewhat less transparent.

One of the most primitive sorts of formal knowledge is that words vary in their *acoustic* duration. The word *mow* has a shorter acoustic duration than the word *motorcycle*. Rozin, Bressman, and Taft (1974) developed a test that they called the *mow-motorcycle* test. It requires the child to indicate which of two printed words, *mow* or *motorcycle*, is the spoken word "motorcycle" (or which is the spoken word "mow") after the examiner and the child pronounce both words. Even a child who cannot read could perform this task on the basis of matching acoustic length with written length. The child who can do this is demonstrating an awareness that print maps sound in some way. However, many preschool and kindergarten children do not perform this test correctly. Lundberg and Torneus (1978) found that Swedish children, who do not begin reading instruction until age 8, did not perform consistently on the *mow-motorcycle* test even at age 6. Their problems in choosing on the basis of acoustic "size" were especially notable when the short word referred to a bigger object than the long word (e.g., *tall*, referring to a pine tree, vs. *tennis racket*.) For some children, semantically big words should be graphemically big words. It would be a mistake to conclude from such a study that children are not aware of print-speech mappings prior to learning to read (although they may be). However, it might be appropriate to suggest that as long as semantic strategies are available to apply to language, formal knowledge will remain underdeveloped or at least less explicit than otherwise.

If preliterate children have difficulty appreciating simple length mappings between speech and print, it comes as no surprise that they also lack explicit phoneme knowledge. Liberman, Shankweiler, Fischer, and Carter (1974) demonstrated that phoneme segmentation is particularly difficult for preschool and kindergarten children. Children were trained to tap once with a stick for each sound (phoneme) in a short word. Thus, *he* gets two taps and *dog* gets three taps. Following training, only a small number of children in kindergarten (age 5) and none in preschool (age 4) could perform this task. Most first-grade children could perform successfully when the task required tapping for syllables instead of phonemes, and all children performed much better for syllables. Thus, whatever difficulties might exist in demonstrating acoustic awareness, phonemic awareness is understandably even more difficult.

It is also clear that explicit phonemic knowledge is related to early reading achievement. Liberman and Shankweiler (1979) report a number of studies that show a relationship between performance on the tapping task and reading achievement. For example, one study (Helfgott, 1976) found that the ability to phonemically segment words in kindergarten was the best predictor of first-grade reading ability, as measured by the Wide Range Achievement Test (WRAT), a test of word reading. Other research shows a similar relation between reading achievement and explicit knowledge of phonemes over a variety of different phoneme tasks (Fox & Routh, 1976; Trieman, 1976; Trieman & Baron, 1982).

The segmenting of speech sounds, even implicitly, is perhaps not readily available to young children. Trieman and Baron (1982) suggested that young children represent spoken syllables as wholes. The structure of syllables as an ordered set of phonemes is not accessible. They found that when subjects were given syllables to classify, adults tended to classify on the basis of shared phonemes but kindergarten children tended to classify on the basis of overall sound similarity. For example, consider the three syllables *bih*, *veh*, and *bo* (pronounced respectively as /bI/, /vɛ/, and /bo/). To decide which two of these syllables sound alike, a subject may pair two on the basis of a shared phoneme, *bih* and *bo*. Or the subject may pair two on the basis of their overall similarity, taking into account the vowel sounds. On the latter basis, *bih* and *veh* have the greatest overall similarity. Trieman and Baron found that children used phonemic similarity less than overall similarity; they tended to group *bih* and *veh* together. Adults use the shared phoneme more often. Trieman and Breaux (1982) have extended this observation to the case of memory confusions. Memory confusions of preliterate children are more likely to be based on overall similarity, whereas those of adults reflect shared phonemes. It seems clear that phonemes are less accessible to consciousness in preliterate children.

Somewhat more surprising, perhaps, is an association between *word* awareness and learning to read. In fact, prior to learning to read, children appear to have little knowledge of printed words and the function of spaces between words (Downing, 1970). Moreover, a number of studies suggest that prereaders do not easily discriminate spoken words from nonwords (Downing & Oliver, 1973–1974; Ehri, 1979). It seems critical for learning to read that the child appreciate that spoken language has semantic units with independent existence (words). In that sense, it is reasonable to say that word awareness is prerequisite for learning to read (Ryan, 1978). It is true, however, that in natural speech, boundaries between words are not consistently marked by silence. Thus our perceptual evidence for words might not be easily accumulated. Indeed, Ehri (1979) has argued that learning to read actually produces word awareness. Although this is probably too strong a claim, it is likely that the child's consciousness concerning lexicality and other linguistic concepts is strengthened by learning to read.[4] In all of this, there is the question of what produces what: Does linguistic awareness produce "readiness" for reading? Or does learning to read produce linguistic awareness? It may turn out that both are true.

Reading at the Beginning Without Linguistic Knowledge

It is possible that the causal relationship between knowledge of phonemes and learning to read runs both ways. The main reason for assuming that phonemic knowledge enables ("causes") reading, rather than vice versa, is a logical one: Application of the alphabetic principle depends on phonetic knowledge. Since letters correspond to phonemes, learning the correspondence logically entails that the learner "know" about phonemes. However, it is possible to begin reading without the alphabetic principle. Or at least it is possible to begin with only a very imperfect application of the principle. Indeed, learning to read Chinese may involve no alphabetic principle at all. Is it not possible that even a child who *eventually* learns to read an alphabetic system such as English may *begin* in the Chinese fashion? That is, might not a child learn by first associating something (a spoken word or a meaning) with a printed word form?

Although the analogy with Chinese is not really a good one, it is possible that the child learning to read in English may bypass the alphabetic principle. In fact, Gough and Hillinger (1980) have claimed that learning to read typically occurs in two phases, associative and alphabetic. In the associative phase the child learns by associating some word from oral language with some print stimulus. The key principles in this phase are that learning is by paired association and that the critical feature of the stimulus, the printed word, is unknown and variable. As Gough and Hillinger (1980) observe, a child may learn to read as his first two words *Budweiser* and *stop*. The word *Budweiser* appears on beer labels and in advertisements with many cues, including a distinctive typography. Perhaps the length of the word is a sufficient cue. The test is to change its appearance in several ways and discover whether the child can still read it. The same test might be applied with *stop,* which might initially be connected with distinctive features of a stop sign. The general principle is that the child can "read" only by association; the cues sufficient for reading responses will vary from word to word.

The child can, according to Gough and Hillinger, acquire a significant vocabulary this way. (We know the Chinese reader can acquire several thousand characters in such fashion.) However, the memory burden for such vocabulary becomes rather large, and after all, the child is reading in an alphabetic system. Eventually, because of good luck, good instruction, or native intelligence, the child catches on to the wonderful riddle of the alphabetic principle. The letters stand for sounds. This is the second phase, the true reading phase.

If this idea is correct in general, or even if it is wrong in drawing a sharp distinction between phases, it provides a most general understanding of learning how to read. Learning to read is learning associations between print stimuli and oral language responses. What features of the stimulus control the response is variable and part of the learning process. At the beginning, it may be very idiosyncratic and may appear (deceptively) to be "holistic." Later, it may be more systematic. Moreover, it is also likely that different features compete for "stimulus control" during early and intermediate phases of learning.

Letters, word shapes, and phonemes
Among these features, the visual shape of the word and its constituent letters are the most important that are intrinsic to the word. An important extrinsic feature is the meaning context that influences the child's expectations about the word. These expectations play a prominent role *before* the child begins to master the intrinsic word features. Thus children's errors in oral reading show a change over the first year of reading instruction. At first, errors tend to be appropriate for the context but often show little graphemic resemblance to the printed word. By the end of the first grade, the errors show a large degree of graphemic overlap (Biemiller, 1970; Weber, 1970). Thus the beginning reader quickly learns to allow intrinsic visual features to control the reading process.

During this beginning period, both constituent letters and word shapes are important features. Especially important is the first letter of a word. For example, when first-grade children have to decide which of several alternatives look most like a nonword they have just seen, they choose the alternative that shares the first letter over alternatives that share other letters (Marchbanks & Levin, 1965; Williams, Blumberg, & Williams, 1970). Kindergarten children, however, show a less consistent use of the first letter in this task (Rayner & Hagelberg, 1975; Williams et al., 1970). Thus the beginning reader is learning, at least partly through left-right visual scanning procedures, to attend to constituent letters and especially the first letter.

As for shape, a distinction is necessary between global shape and specific shape, which includes letter features. For example the letter string *cug* is more similar to *arp* than to *tuk* in global shape; *cug* and *arp* both have descenders at the end and no ascenders or descenders at the beginning. However, such global shape cues have very little effect on children's choices when they have to decide which alternative looks like a letter string they have just seen (Marchbanks & Levin, 1965). In fact, in the example above, if they have seen *cug*, first-grade children will be more likely to chose *tuk*, which has only a medial letter in common with *cug*, then *arp*, which has global shape in common. However, if word shape is defined to take into account the distinctive features of letters as well as global shape (Gibson, 1969), the situation is different. For example, *cug* is more similar to *cwq* than to *cqn* and more similar to *ouq* than to *jun*. That is, *w* is more similar to *u* than *q* is, and *o* is more similar to *c* than *j* is in terms of distinctive features. Under these conditions, Rayner and Hagelberg (1975) found that first-grade children chose from the alternatives based on shape as well as letter. That is, after *cug* is shown and removed, subjects were more likely to choose *cwq* than *cqn* and more likely to choose *ouq* than *jun*. Rayner (1976) found the use of shape cues to increase with age from kindergarten, where they are not used, through adulthood, where they are used even more than letter cues.

A study by Beverley (Beverley & Perfetti, 1983) has examined a similar issue: Are children learning to read sensitive mainly to the visual and graphemic features of printed words or are they also sensitive to phonemic information? In this study, second-grade readers were compared with older children and adults on their judgments of word similarity. Sixty-six pairs of printed words were presented for children's judgments of similarity according to whatever criteria they chose to use. The visual, graphemic, and phonemic similarity of the pairs varied. For example, *new* and *sew* are identical in length and similar in shape (visual features) and share constituent letters (graphemic features) but are different phonemically. *Sew* and *show* on the other hand, are not similar in length and shape but they have some grapheme similarity and very high phonemic similarity. Given pairs that could vary in the similarity of visual, graphemic, and phonemic features, the question was what kinds of features would be used to judge overall similarity. For second-graders, shared visual features (word length), shared letters, and shared phonemes all contributed to perceived similarity, with shared letters receiving the most weight. In fact, shared letters, independent of shared phonemes, were the most important feature for all age groups, including adults. The weight given to shared phonemes depended on age and reading skill. Adults and skilled readers in the fourth and sixth grades used shared phonemes as an important feature, but second-grade subjects and less skilled fourth-grade subjects did not use phonemic similarity. Second-grade subjects were especially unlikely to use vowel similarity as a feature. One reasonable interpretation of these results, although not the only one, is that younger readers are less tuned to the way a printed word "sounds" than to the way it "looks."

The results of these experiments suggest that the course of reading in the early grades involves learning features of print and using these features in fairly intricate ways. Global visual features, which may serve a prereading stage along with other cues, lose value very quickly. The constituent letters themselves dominate early reading processes, both in their specific identities and in their contribution to shape features of the word. Finally, the phonemic values of the letters may play a smaller role than their intrinsic visual identities.

Letter names and phonemes

Children tend to learn the alphabet as part of their prereading experience, and something from this alphabet learning is probably applied to word reading. The alphabet learning experience has two components: the perceptual learning of the letter forms and the associative learning of letter names. Both of these can have some role in learning how to read. The perceptual learning of letter forms means that words will have familiar parts, even if the child does not analyze a word as a string of letter constituents. The letter name learning may help attach linguistic significance to letters, even without the child knowing the alphabetic principle.

In fact, there has been much controversy and a bit of research about whether learning the names of the letters helps in learning to read. Although it has often been concluded that learning the names of the letters does not help, it appears that this conclusion is incorrect. Ehri (1983a) has carefully examined the evidence from studies in which some children were taught letter names and other children were not. Her conclusion is that learning letter names does help, or at least it does not hurt.

One reason for the argument against letter name learning has been logical. The names of the letters are not equivalent to the phonemic values associated with the letters in the writing system; "dee" is not the phonemic value of *D* and "double you" is far from the phonemic value of *W*. However, as Ehri points out, the important thing may be not learning of correct phonemic values, but learning some speech-based association to the letter form. Furthermore, the letter names typically do include some of the relevant phonemic values. There is a /d/ in "dee" and there is an /s/ in "ess." Thus, although the link is very imperfect, there is useful phonetic information in the name of a letter and very young children demonstrate that they can use this information. Examples from Read (1971) and Chomsky (1977) show that spellings invented by beginning readers and even prereaders reflect phonemic information in letter names. For example, children between 4 and 6 who know the names of the letters but who cannot yet read invent spellings that are phonemic compromises: KAM (*came*), FES (*fish*), WOT (won't) and JMEZ (Jimmies) (Chomsky, 1977). Interestingly, vowels are often omitted with consonants whose names prominently include a vowel sound (e.g., *R* and *L*) resulting in, for example, TIGR, GRL, and KLR (color).

Given evidence for such phonetic-orthographic creativity among very young children, Chomsky (1971, 1977) suggested that invented spelling provides a good introduction to reading. Whether or not invented spelling is to be advocated, the main point for the alphabet is this: Letter names may not be a serious obstacle to learning to read. An ideal system would be one that associates phonemes with letters, that is, one in which their names are their phonemic value. Since this is impossible for English because of the many multivalued letters, the choices are reduced to two: Teach letter names before learning to read, or postpone letter names until later, after the child has learned the phonemic significance of letters. It is quite possible that either method might be satisfactory, provided it gives the child a linguistic association between speech and print.

Ehri (1983a) makes a similar point about letters and letter names. In addition, Ehri emphasizes the importance of print for a visually based memory for words and phonemes (Ehri, 1978, 1980, 1983b). Ehri's theory is that spellings of words serve as a memory representation as part of an "orthographic image." In the case of word spellings, the orthographic image can guide the pronunciation of a word as well as provide part of its overall memory representation. In the case of a single letter, it provides a unique symbol for a speech sound. Ehri (1983b), in fact, sees the relationship between phonemic awareness and learning to read not in terms of awareness leading to reading, but in terms of the symbol value of letters. By her theory, letters allow the child something to associate phonemes to. Only after letters are learned can the child direct attention to meaningless sounds. In fact, experiments support the conclusion that phoneme segmentation performed by prereaders on spoken syllables is facilitated if they have been trained to use letters to symbolize the sounds. For example, if a child is asked to segment the spoken syllable *man* by marking each phoneme with a token, performance is better if the child has first learned to perform the task by using the letters *M, A, N* to mark each segment. The implication seems to be that, despite the imperfect mapping of letter names to speech sound, the names can help raise awareness of phonemes. The visual letter form, even with its name, can provide a symbol for representing phonemes.

Summary

To this point, we have emphasized these critical observations about learning to read in an alphabetic language: (1) It depends ultimately on application of the alphabetic principle. (2) The application of the alphabetic principle depends in part on knowledge of phonemes as meaningless units of speech, but this knowledge is not accessible to many prereaders. (3) It is possible, even likely, that initial progress in reading can occur without application of the phonetic principle, and hence, without explicit phoneme knowledge. Useful phonemic knowledge itself may be nurtured by alphabet learning. (4) Beginning readers are tuned into the visual and graphemic features of words using especially initial letters to cue their lexical access.

A Reciprocal Relationship Between Phonemic Knowledge and Learning to Read: A Longitudinal Study

One implication of the preceding discussion is that the relationship between phonemic knowledge and learning to read is not simply that the first causes the second.[5] The general nature of the relationship may be reciprocal: Knowledge of phonemes enables learning to read, and learning enables phoneme knowledge. There is evidence that learning to read can come first. Morais, Cary, Alegria, and Bertelson (1979) found that a group of Portuguese adults who had not learned to read could not perform the phoneme analysis task of Liberman et al. (1974). However, a group of adults who were comparable to the illiterate group except for just having been through an adult literacy program could perform the task. The reasonable conclusion is that the phonemic knowledge tapped by this task does not develop spontaneously, but rather depends in part on reading experience. This conclusion, of course, cannot be applied to children who might develop some phonemic knowledge independent of instruction. The potential for demonstrating knowledge on school-like tasks, even speech-based ones, probably declines in the absence of schooling. Nevertheless, this is the kind of demonstration that points to an important role for reading in the development of phonemic knowledge.

The longituinal study of Perfetti, Beck, and Hughes (1981) was designed to examine the progress of children through the first grade on both phonemic knowledge and learning to read. The assumption was that a reciprocal relationship would be found: Learning to read would partly depend on phonemic knowledge and vice versa.

Two kinds of phonemic knowledge,
The reason for mutual influence is that phonemic knowledge is not unitary. One kind is knowledge of phoneme *synthesis*. It depends on recognizing that a meaningless sound spoken in isolation is a speech unit that can combine with other such units to form words and syllables. For example, if the three phonemes /k/, /ae/, /t/ are spoken in isolation, they can be synthesized into the word *cat*. A child who cannot make such a synthesis may lack the kind of phonemic knowledge necessary for reading. Notice that this knowledge does not entail conscious manipulation

of phonemes as abstract objects. It does require the insight that *meaning-less sounds* add up to meaningful words (and meaningless syllables).

The second kind of knowledge is phoneme *analysis* (or, more care-fully, *syllable analysis*). This is the more explicit knowledge that sylla-bles and words are made up of meaningess constituents. The explicit-ness of the knowledge is demonstrated by the child's manipulation of the phonemes. In the Perfetti et al. (1981) study, this manipulation was achieved on a task adapted from Rosner (1973) and Bruce (1964). The child is asked to delete a phoneme from a spoken word and to produce the result of the deletion. For example, the child first says the word *cat* and then is asked to say "cat without the /k/" or "cat without the /t/." Only initial or final consonants, never medial vowels, are deleted. The result is sometimes a word but often not. Notice that knowledge needed to perform this task is more explicit and more analytic than that required by the synthesis task. It is more explicit because it requires the child to perform a task that can be done only by noticing and consciously pro-ducing phonemes. (In the case of a two-segment word such as *on*, the child must actually produce an isolated segment.) It is more analytic be-cause it depends on examining a word or syllable for its constituent parts. This is a kind of knowledge that presumably depends on, rather than pre-cedes, learning to read. By our account, it is promoted by the child's en-counters with printed words, since these, perhaps for the first time, may allow the child to notice that words have parts.

Procedures of the longitudinal study

The tasks described above, synthesis and analysis, were given, to 99 be-ginning readers, 82 of whom were true beginners and 17 of whom were repeaters who had failed to learn to read. The 82 beginning readers formed three groups from two different schools.

One group was a *direct code* group who were being taught by explicit direct instruction in the alphabetic principle. The program, known as NRS (New Reading System) and developed by Isabel Beck (see Beck & Mitroff, 1972), directly teaches correspondences between letters and phonemes. It also teaches "blending," a procedure for combining word sounds as they are learned. (In NRS, the blending procedure uses visual prompts, so that the child sees letters moved together as the phoneme blend is spoken.) Thus, the teaching method for the direct code subjects was a direct code-breaking approach that included grapheme-phoneme pairs taught in isolation and synthetic formation of words from their parts.

The other two groups were from a different school that taught reading with a widely used basal reading series. As the result of performance on reading readiness tests taken at the end of kindergarten, students either began first-grade reading in the first reader of the basal series or else continued in the "reading readiness" workbook begun in kindergarten. The first group (called the "Rockets") obviously constituted a more advanced group than the second (the readiness group).

The interest in the three different subject groups was in how the method of instruction was related to phonemic knowledge, learning to read, and the relationship between the two. One basic question is whether a group whose instruction depends on letter-phoneme correspondences (direct group) is more dependent on phonemic knowledge than a group that does not receive such instruction (the basal groups). Each student was tested four times during the school year. The first testing began on the second day of school, prior to any reading instruction (Test T_0). Subsequent tests followed at the end of October (T_1), mid-January (T_2), and the end of April (T_3). Each test involved a check of the child's ability to read orthographically regular pseudowords; at the end of the year, the child's general word-reading ability was tested by the WRAT. The test of pseudoword reading may be considered a test of true decoding.

Phonemic performance over the year

Figure 10-1 shows the percentage of subjects achieving criterion on the synthesis task. Few children could perform synthesis prior to instruction, but most mastered the task before the end of the year. In fact 100% of the direct code group had passed by the end of the year. At the eight-week point, however, only 24% of this group had reached criterion, despite the fact they were receiving grapheme-phoneme instruction and blending instruction. Even though these students were required to synthesize with letters in their curriculum, they did not yet transfer this synthesis skill to isolated phonemes without letters. However, they, more consistently than any other group, did eventually catch on to phoneme synthesis.

Figure,10-2 shows the very different results for phoneme deletion. Performance was generally lower, except that nearly one in five of the advanced basal group (the rockets) could perform the task prior to reading instruction. However, progress during the first eight weeks of instruction. Dramatic gains occurred between the 19th week and 33rd week for the direct code students and between the 8th and 19th week for the Rockets. The lower basal students did more poorly, with only about one in four reaching criterion by the end of the year.

Reading progress and phonemic abilities

Of central interest is the relationship between the child's developing reading ability and his ability in phoneme synthesis and analysis. For the two basal groups combined, the best predictor of a child's year-end reading achievement (WRAT) was performance on the deletion task. However, it was a better predictor after instruction ($r = .73$ at T_1 and $r = .79$

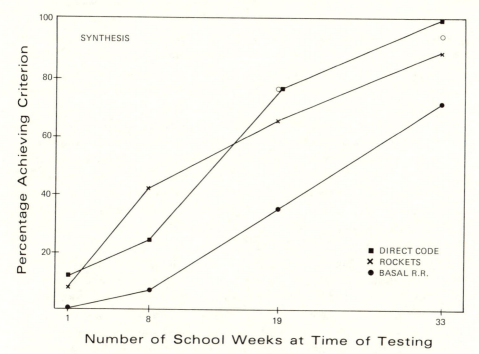

Figure 10-1. Percentage of subjects achieving criterion on phoneme synthesis at each test point (Perfetti, Beck, & Hughes, 1981).

Figure 10-2. Percentage of subjects achieving criterion on phoneme deletion at each test point (Perfetti, Beck, & Hughes, 1981).

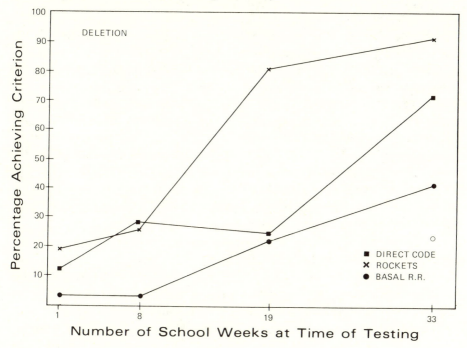

at T_2) than prior to instruction ($r = .59$ at T_0). Prior to instruction, synthesis was a good predictor of year-end achievement ($r = .62$), but it was only slightly better than deletion. Unlike deletion, synthesis did not improve in predictive power over the year.

For the direct code group, the situation was both similar and different. Direct code subjects were similar to the basal group in the predictive value of deletion. Deletion measured prior to instruction was a moderately good predictor of year-end reading achievement ($r = .43$),and increased its predictive value at later test points ($r = .72$ at T_1 and T_2). The difference was in synthesis, which was never a very good predictor of year-end reading achievement. Prior to instruction, its correlation with year-end reading achievement was negligible ($r = .21$ at T_0) and did not increase until the last test point in April ($r = .47$ at T_4).

Why are there these similarities and diferences between the groups? The explanation is as follows: Phoneme analysis is a fair predictor of year-end achievement regardless of instructional method. Its predictive power increases when it is measured later because performance on the experimental task and on the year-end reading achievement test both benefit from learning how to read. Synthesis is a more primitive linguistic ability than phoneme analysis. It is a good predictor of achievement prior to instruction for students who will not be explicitly taught relevant phonemic principles (i.e., basal students). For direct code students, synthesis is less predictive of year-end achievement because these students will be taught explicitly the relevant phonemic principles.

It is possible to examine more carefully the causal relationship between phonemic knowledge and learning to read. Because of the repeated testing, two questions can be answered: (1) What is the relationship between the test point at which the child first passed criterion on a phonemic task and his performance on the final reading achievement test (WRAT)? (2) How does decoding performance, as measured on the pseudoword tests, compare at test points before and after the child passed criterion on a phonemic task?

With respect to the first question, basal students who passed synthesis early scored higher on the year-end achievement test than did subjects who passed it late or not at all. A way of illustrating this is shown in Figure 10-3. It is a tree diagram branching on the left those subjects who passed the synthesis test at each test point, and on the right those who did not pass. At the bottom of each branch is the mean year-end achievement score of subjects on that branch. For example the 2 subjects who passed criterion prior to instruction (T_0) eventually had a mean reading score of 65.5 (WRAT raw score) and the 18 subjects who passed at the last test point had a mean score of 42.4. The time at which criterion was achieved is highly related to year-end achievement. In practical terms, students who reached synthesis criterion at T_1 were nearly one grade level

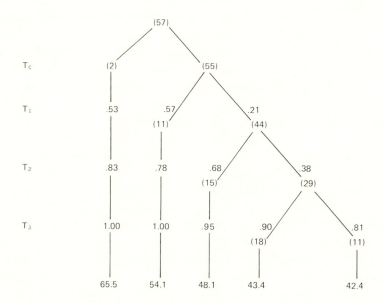

Synthesis
Basal Group
(n = 57)

Mean Reading Scores (WRAT) at T₃

Figure 10-3. Tree diagram showing end-of-year reading scores (WRAT raw score) of 57 basal subjects. At the bottom of each branch is the mean reading score of subjects who passed the synthesis criterion at a given test point. The number of subjects who passed criterion is in parentheses at the node of each left branch. Also shown at each test point is the decoding score as a percentage on the pseudoword reading task. For example, the eleven subjects who achieved criterion at T_1 had a decoding score of .57 at T_1 and a score of .78 at T_2 (Perfetti, Beck, & Hughes, 1981).

higher in year end achievement than those who passed at T_2, some 11 weeks later.

Deletion showed a similar pattern: The earlier a student reached criterion on deletion, the higher the year-end achievement score.

For direct code students, the picture is slightly different. All students passed the synthesis test eventually. Students who passed the synthesis test early had a slightly higher year-end reading score, but the difference was not dramatic. For example, the four subjects who passed at T_1 were only about four points higher than the five subjects who passed at T_3. Again the conclusion is that year-end achievement is not highly dependent on early phoneme knowledge when the relevant knowledge is going to be taught.

A different picture emerges when performance on the pseudoword tests is examined. Since these tests were given at each test point, progress in learning to decode novel word forms can be assessed. For direct code students, performance on pseudoword reading *increased* at the same test point at which phoneme synthesis was passed. Furthermore, pseudoword reading generally increased again at the test point immediately after synthesis was passed. Importantly, the same pattern held for the basal subjects. Thus, irrespective of instruction, progress in learning to decode novel word forms is linked to the ability to synthesize phonemes.

This brings us back to the question of what causes what in the relationship between learning to read and phoneme knowledge. The fact that there is a correlation between year-end achievement and any particular phoneme test does not help much in deciding. However, if we look at partial time lag correlations between the phoneme tasks and the pseudoword reading tasks, we can address whether progress in phoneme knowledge precedes or follows progress in decoding (pseudoword reading). The answer is both. The pattern of time lag correlations shows that progress in phoneme *synthesis precedes* progress in pseudoword reading for both basal and direct code subjects. Progress in phoneme *analysis* (deletion) *follows* progress in pseudoword reading for both. The only difference between the two groups is that basal subjects continue to show gains in reading following gains in phoneme deletion. This state of affairs is depicted in Figure 10-4. The arrows indicate that a score on a given test significantly predicts a score on a subsequent test even when all other correlations are taken into account.[6]

Conclusion

The results of the longitudinal study suggest that learning to read partly depends upon phonemic knowledge and partly produces it. Both synthesis and analysis of phonemes predict year-end success in reading words. However, the reasons for their predictive values may be somewhat different. Synthesis is an *implicit* speech segmentation ability related to decoding. Success in novel word decoding depends on this ability. Analysis is an *explicit* speech segmentation ability that is not necessary for at least some decoding. Learning to decode creates the conditions for developing explicit phoneme knowledge.

Summary

In an alphabetic language learning to read requires, ultimately, application of the alphabetic principle. It is possible, however, for reading to begin without clear access to this principle. Indeed, it is clear that preliterate children do not have easy access to phonemes. Thus, reading for

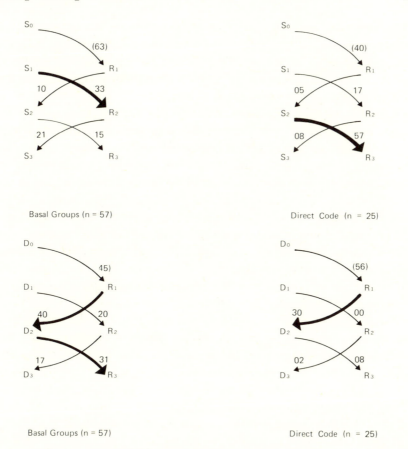

Basal Groups (n = 57)   Direct Code (n = 25)

Basal Groups (n = 57)   Direct Code (n = 25)

Figure 10-4. The relationship between phoneme synthesis and pseudoword reading (top panel) and between phoneme deletion and pseudoword reading (bottom panel). Darkened arrows indicate significant time lag correlations with same-time correlations partialed out. For example for the basal group, synthesis at T_1 predicts pseudoword reading at T_2. Pseudoword reading does not significantly predict synthesis on a later test. It does predict deletion on a later test for basal subjects (lower panel) (Perfetti, Beck, & Hughes, 1981).

most children begins with at best imperfect knowledge of the structure of speech. The experience of learning to read itself helps produce this knowledge.

ABILITY DIFFERENCES IN EARLY READING

In earlier chapters, we discussed ability differences in reading comprehension by reference to basic processes of word coding. It is likely that the genesis of such ability differences is in beginning reading. Indeed, individual differences in word reading are necessarily very large during beginning reading.

The essential reading ability observable among readers is oral word reading. Reading words is in fact the standard of ability, as reflected in the Wide Range Achievement Test (WRAT), a test of word reading. If application of the alphabetic principle is eventually the key to reading competence, then it is possible to assume that ability differences lie in the knowledge and procedures children have for application of this principle. For example, we have seen that explicit linguistic knowledge is part of an intricate reciprocal relationship with decoding. Children who manage some progress in learning to read, but who remain low in ability during the first or second grade, may have missed out on some of this knowledge.

On the other hand, it is clear that the full range of ability factors is relevant even for beginning reading. For example, differences in vocabulary are already quite large by age 6. So too are related differences in verbal and nonverbal intelligence. Furthermore, there are general cognitive development factors that are important for beginning reading. Such factors cannot be ignored altogether. One key to a young reader's achieving the potential allowed by other factors is learning the coding system. High-ability readers in the first or second grade are children who can read words and even nonwords. (This is not a question of the teaching of reading. Many children learn to read before they start school [Durkin, 1966; Torrey, 1979].)

Although early differences in reading ability prominently include word coding, the consequences of this fact are just beginning to emerge. Some aspects of this phenomenon were examined in a longitudinal study of children learning how to read (Lesgold & Curtis, 1981; Lesgold & Resnick, 1982). The study is not only informative about ability differences in coding, it also has implications for reading instruction.

The Pittsburgh Longitudinal Study of Beginning Reading

The longitudinal study followed the reading progress of a group of children from the beginning of first grade through third grade. The children were a racially mixed group from a Pittsburgh "urban suburb."

The ability issue was addressed by grouping subjects into three skill levels based on second- and third-grade comprehension scores. There was a clear separation of a low-ability group and a medium group by the second year, with the low group averaging below the 30th percentile and the medium group averaging just above the 50th percentile. One of the intriguing findings of the study was that these two groups were not different in their comprehension test scores at the end of first grade. The lowest ability group *was* lower on tests of vocabulary and phoneme-grapheme knowledge in first grade, however. Differences between low-

and medium-ability readers on these tests actually increased between first and second grade. Since these differences are based on nationally standardized test scores, it is fair to summarize the ability differences as follows. Children not successful in learning to read began with low skill in certain tests of basic skills required for reading. They failed to improve in these skills over time, falling farther behind their successful peers. Their problems in comprehension did not show up until the second grade (Lesgold & Resnick, 1982).

The longitudinal character of the study comes from tests given as children completed curriculum units (a basal curriculum). The low-ability children generally made slower progress through the curriculum, although a few children who were low in ability, as measured by the standardized tests, were placed by teachers in faster groups. Since even these children did poorly in comprehension tests later on, Lesgold and Resnick rhetorically posed an interesting practical question: What skills do such children have that allow them to move quickly through basal readers while masking reading deficiencies that show up in other situations?

Oral reading
Of major interest is performance on certain tasks given at the curriculum points to assess word coding speed and oral reading, among other things.

Children read passages based on the words and sentence patterns of their readers and passages designed from different materials to test transfer. Low-skill children made more errors than medium- and high-skill children in both sorts of passages. (Differences between medium- and high-ability groups were more pronounced on the transfer passages.)

Errors made in oral reading were also examined according to a classification scheme described by Hood (1976). One important result concerns the phonemic overlap between a child's error and the target word on the page. The lower the level of skill, the less often was there more than 50% overlap in phonemes. A second striking result was the tendency for low-skill readers to skip words that were difficult. Finally, when the errors were scored for appropriateness, the vast majority of substitution errors of all skill groups were appropriate for the context. However, low- and medium-skill readers made *fewer* inappropriate substitutions than high-skill readers. On the other hand, low-ability readers made *more* errors that were only "partly appropriate" as opposed to "completely appropriate." The picture that emerges is one of a low-ability reader with weak coding ability. Such a reader's strategy is to do as well as possible in the absence of this skill. He or she uses context and makes lots of contextually appropriate errors, sometimes producing words that are very dissimilar in structure to the word on the page and sometimes avoiding the word altogether.

Verbal efficiency

The hypothesis that reading comprehension depends partly on verbal efficiency is also addressed by the longitudinal study. Students were tested at each point on speed of word recognition, measured by vocalization latency to single isolated words. These measures were combined with speed of oral reading to constitute a speed-of-processing variable. Since there were four tests of vocalization speed and oral reading and three measures of standard reading comprehension, it was possible to ask whether *speed* preceded *comprehension* more than vice versa. That is, would a student's speed measures at Time 1 predict comprehension at Time 2 better than vice versa? This is analogous to the question posed by the longitudinal study of phonemic knowledge described earlier in this chapter. The result reported by Lesgold and Resnick (1982) suggests that speed predicted comprehension. In short, a child's first-grade speed score was a very good predictor of that child's second-grade comprehension score. There were no significant links running in the other direction.

This is not to say that links from earlier comprehension to later processing speed might not exist. Different measures of comprehension or more frequent measures of comprehension may give a better opportunity for comprehension-to-speed links. Given this qualification, the conclusion of Lesgold and Resnick (1982) seems appropriate: "The overall pattern of results provides confirmation of the assertion that automaticity is a cause of more adequate overall reading skill and not an epiphenomenon produced by the fact that better readers practice more."

SUMMARY

Ability differences are observed in reading from the beginning of reading instruction and even before. There are several important components to these differences. In learning to read an alphabetic language, a major factor is the abstractness of the phonemes onto which letters are to be mapped. This factor, along with the complex vowel mappings, creates obstacles to learning how to read. These obstacles can be serious for children without adequate linguistic knowledge. Explicit knowledge about speech segments seems to develop along with learning how to read for the successful learner. The relationship between phonemic knowledge and learning to read is best described as reciprocal. The low-ability reader begins reading with less accessible linguistic knowledge and may fail to sufficiently acquire this knowledge during the course of learning to read.

Some ability differences in reading can be masked in the beginning only to emerge as comprehension problems later. The rudiments of verbal ef-

ficiency are seen early in reading. As successful readers acquire the coding system, they advance, with practice at reading, to a stage of facility
that is characterized by speeded word processing. Low-ability children
do not acquire this level of low effort, affecting the level of comprehension they obtain. Evidence suggests that, consistent with verbal efficiency theory, word-processing efficiency leads to better comprehension,
rather than merely being a by-product of comprehension.

NOTES

1. Phonetic compounds are characters that contain a radical, which has semantic value, and a phonetic, which has pronunciation value. These account for about 80% of Chinese characters (Leong, 1973).
2. It is possible that Chinese may have less to gain from this fact, compared
   with other languages. Words in Chinese are largely single syllables. This
   results in a great many homophones (different words that sound alike).
   Although the tone system, added to the syllable structure, largely eliminates ambiguity, it is possible that the ideographic system is much more
   suitable for such a language than it would be for a multisyllabic one.
   Equivalently, an alphabetic system may be less valuable for such a language than for a multisyllabic one.
3. There has been considerable interest in these facts of English orthography and in whether orthographies represent superficial phonemes or
   underlying phonological and morphological structures (Chomsky & Halle,
   1968; Venezky, 1967). It is also clear that there are differences among
   languages in whether their orthographies are deep (morphological) or
   shallow (more phonemic). These differences may have some implications for how words are read (Katz & Feldman, 1981).
4. It is not clear what to make of the claim that prereaders lack word
   awareness. The experiments of Downing and Oliver (1973–1974) seem
   to allow various interpretations of the apparent discrimination problems
   of prereaders, as Ehri (1979) observes. Ehri's (1979) own experiments do
   not really demonstrate that prereaders lack the concept of a word as independent speech object. One study (Ehri, 1976) demonstrated that
   prereaders have difficulty learning word responses in a paired-associates
   task, and another study (Ehri, 1975) demonstrated certain problems in
   locating words as sentences were spoken and in other fairly complex tasks.
   In the most direct and simple test of word awareness, Ehri (1979) reports
   that most children could correctly classify nonsense words as nonwords
   even though they were less successful on *real* words. It appears that, as
   is usually the case with tests of implicit linguistic knowledge, it is very
   difficult to isolate the basic linguistic knowledge from peripheral task
   demands. Actually it is quite likely that young children in some sense
   appreciate some properties of lexicality. They do not hesitate to produce
   isolated words. More significantly, at a very young age they exhibit
   knowledge of morphological rules by adding inflections (e.g., plural forms)
   to *nonwords*. This indicates knowledge of lexical morphemes. What they
   may lack is "word consciousness" as they lack all linguistic conscious-

ness. They have had no occasions to incorporate "wordness" into a conscious structure, and when they are required to perform some complex task they use whatever they can to perform it. The more complex the task, the less likely they are to use "wordness" or any concept not high in consciousness.

5. There is evidence from a training study of Bradley and Bryant (1983) that a causal relationship exists between phonemic knowledge and reading achievement.

6. Figure 10-4 shows the essential results of time lag correlations. These correlations are, for example, the correlation between synthesis at T_1 and pseudoword reading at T_2 with the correlation between synthesis at T_2 partialed out. The reverse pattern was also computed: Pseudoword reading at T_1 and synthesis at T_2 with the correlation of T_2 reading partialed out. With such a procedure, some correlations from one time to the next become nonsignificant because of the contribution of the same test correlations. What remains as significant is what is shown in Figure 10-4.

CHAPTER ELEVEN

Verbal Efficiency and Reading Instruction

The central question addressed in this chapter is whether the theory of verbal efficiency has implications for teaching: Should decoding be emphasized over comprehension instruction? Should linguistic processes, aside from print, be emphasized in instruction? Should remedial training include decoding? The first section considers general implications of verbal efficiency. The following sections discuss beginning reading instruction, comprehension instruction, and remedial instruction.

GENERAL IMPLICATIONS OF VERBAL EFFICIENCY

We begin by recapitulating the major observations that have formed the underpinnings of verbal efficiency. These are listed below in summary form. Previous chapters contain elaborations of each point.

Principles of Reading Ability
General Comprehension. Reading comprehension is the result of processes that operate on many different levels. Prominent among these processes are lexical access, propositional encoding, and construction of mental text models. Reading ability is limited to the extent that limitations in word access, working memory, or relevant knowledge (including text knowledge and conceptual knowledge) are present.

Low Verbal Efficiency. Low ability in reading is almost always accompanied by low-level verbal coding problems. These include especially ineffective or inefficient word decoding, although they often extend to reduced abilities in verbal memory. Thus, the comprehension of texts is limited by verbal coding inefficiency.

High Verbal Efficiency. The hallmark of a skilled reader is the efficient activation of word-level (name) codes from an inactive memory. Such a reader's lexical processes are not dependent on context, although context may be used to build a text model.

Linguistic Processes. Reading is a linguistic process as well as a visual one. The major components of general reading ability are not necessarily unique to print. Verbal memory especially seems to be a very generalized processing factor. Furthermore, implicit speech processes are an important part of skilled reading. These processes are at least part of what skilled verbal coding is about.

Developmental Continuum. There are some continuities in the development of reading skill despite some differences. Children with specific reading disabilities show many of the same verbal processing problems that low-ability adult readers do. Some of these verbal processing factors extend to adult reading ability.

Beginning Reading. For a beginning reader in an alphabetic script, learning the alphabetic principle is a major achievement. This achievement and the equally important ability to apply the principle skillfully are related to knowledge of speech segments in complex ways. Continued progress in mastering decoding, including improvements in efficiency, allows gains in text reading.

Some implications
The main implication of verbal efficiency for instruction in reading is not how to teach but what to teach. It is obvious that children should learn something about decoding. They must learn that elements of the script system, the letters, map meaningless elements of the speech system. Of course, this is a weak conclusion. It has been reached time and again by thoughtful writers (Rozin & Gleitman, 1977; Liberman & Shankweiler, 1979).

Can we go beyond the agreeable conclusion that the child should learn the alphabetic principle? Clearly, a second implication is that the child should learn *enough* about decoding so that words may be read without effort. Slow and accurate should give way to fast and accurate. Thus a general implication of verbal efficiency theory is that efficiency of word access is a useful instructional goal beyond accuracy.

Does this mean that practice at word recognition should be encouraged, so that speed of access increases? Perhaps. But such a conclusion does not follow without some additional assumptions. The main assumption required is that a child has adequate knowledge of decoding principles and lacks only automaticity. The second assumption is that automatic or low-effort decoding can be equated with speed. The first assumption is often not met by children for whom such practice might be recommended. The second assumption is clearly false, but it probably does not matter much. As a practical matter, the correlation of speed with automaticity means that, on the average, increases in one will produce increases in the other.

The adequate-knowledge assumption deserves more attention. Practice at speeded processing (for example, saying printed words fast) makes sense only for a reader whose knowledge is high but whose level of practice is low. Ideally, this describes a child reader who has mastered orthographic principles. Such a child can convert printed words to spoken words and, as a more stringent test, can read orthographically regular pseudowords (such as *slint*) as if they were real words. If the knowledge is not adequate, there is no reason to expect that speeded practice per se will do any good. It would be like expecting a novice chess player to become a better player by making each move within five seconds. The procedure might work, but only if some chess knowledge is acquired as a by-product of the rapid moves. Similarly, a low-knowledge reader might benefit from speeded practice by acquiring knowledge of decoding principles, such as orthographic structure, as a by-product of saying words fast.

The general implication of verbal efficiency is that, since decoding efficiency is instrumental in reading comprehension, it should be mastered. It is clear, however, that lexical access is a fairly complex business and that children who are not good at it typically have insufficient knowledge concerning orthography or related components of lexical knowledge. This leads to a different perspective: Children need to acquire a rich word representation system, including the decoding principles. Practice will increase efficiency given adequate knowledge, but speed of word processing itself is a by-product of the acquisition of lexical knowledge and of practice at using this knowledge. Thus speed of processing reflects knowledge and practice, and lack of speed reflects lack of knowledge or lack of practice. Speed is not the cause of anything, merely a consequence of overlearning.

This is not to deny that speed of processing is important. Its importance lies in telling us something about the verbal processing of a child. Slow processing tells us there is a problem in the child's decoding, but it does not point to speed as the solution to the problem. It is quite possible that teaching of decoding principles will be helpful for such a child.

It is, in fact, probable that children of low ability have not mastered decoding principles. This is true of children in, say, the fourth and fifth grades. It is also probably true of adults who do not read well. Curtis

(1981), in her study of college students' vocabulary knowledge, found that low-knowledge subjects had problems simply reading words, in addition to their problems with meanings. Thus teaching decoding principles may be a useful objective for younger and older readers. This is not to say, of course, that direct and explicit teaching of decoding is desirable for adults—only that learning decoding principles may be necessary.

There is a second incorrect inference that is sometimes drawn from verbal efficiency. Because of the emphasis on decoding, verbal efficiency may seem counter to the teaching of comprehension. It is true that decoding is central and that instruction should be directed at decoding. But it does not follow that comprehension instruction is not helpful. Indeed, the interactive framework clearly assumes that other comprehension processes can be responsible for comprehension difficulty. Furthermore, weaknesses in decoding may be compensated for, to some extent, by other comprehension processes. A reader's word-identification ability may be low, but ability to use context may compensate for this weakness. Stanovich's (1980) interactive compensatory model, like the verbal efficiency model, makes this especially clear. However, comprehension instruction can do more than merely compensate for weak decoding abilities. Children who have difficulty in reading have problems other than decoding. They tend to have generally poor linguistic processes, a lack of knowledge, and ineffective comprehension-monitoring skills. Many children can use all the help they can get. Verbal efficiency claims that comprehension will not reach its potential until word coding is efficient. But it does not deny that other improvements in comprehension are possible.

These are the general implications of verbal efficiency theory. They do not say anything about how to teach reading—merely that decoding should be learned somehow. Both speed and knowledge of decoding principles are objectives of instruction. Comprehension instruction can also raise comprehension levels, although decoding efficiency may remain an obstacle.

BEGINNING READING INSTRUCTION

The teaching of reading has had a long history of contention. Although the argument has taken different forms, the central point of contention has been decoding—whether to teach it directly, whether to teach it early or late, and perhaps whether to teach it at all. The major alternatives to instruction programs that emphasize decoding have been programs that emphasize meaning. The "whole-word" or "look-say" methods have been widely used alternatives to decoding-emphasis programs. In fact, the whole-word method was the predominant method of teaching reading in the United States at least until recently, and meaning-emphasis programs remain among the most widely used reading programs.

There have been general shifts toward including "phonics" instruction within meaning-emphasis programs. However, these programs devote less attention to what gets taught in their phonics and what gets read in their stories (Beck, 1981). The meaning-emphasis programs rely mainly on stories (the "basals") and the building up of a sight vocabulary. This fact is reflected in the choice of vocabulary that occurs in the meaning-based or basal programs compared with the code-emphasis programs. In a study comparing the content of basal readers with code programs, Willows, Borwick, and Hayvren (1981) found dramatic differences in the first 50 words introduced by the programs. The code-emphasis programs introduced words that were orthographically regular and that included a more restricted set of grapheme-phoneme correspondences. The basal programs had a less selective set of words, both more irregular words and less controlled correspondences. The principle of the basal readers is to introduce words familiar to the child from oral language experience. The principle of code programs is to teach grapheme-phoneme correspondences, even though this sometimes results in lessons that contain words of lower frequency (Beck, 1981; Willows et al., 1981).

The gain that basals may achieve in word familiarity tends to be offset by losses in decodability. Beck (1981) made a comparison of four code programs with four basal programs. In the basal programs, she found practically no words in the stories read during the first one-third of the year for which all the grapheme correspondences were taught. Even by the end of the year, the basal programs had taught correspondences sufficient for only about half the words appearing in stories. Code-emphasis programs, by design, controlled vocabulary so that from 79% to 100% of the words, depending on the program, were always decodable on the basis of taught correspondences. The point is not that the code-emphasis programs are pedagogically superior; that is a more complex issue. But it is clear that the basal readers, as Beck observed, make it "more difficult for beginning readers to recognize that certain letters have certain sounds, and that words encountered in reading can be decoded on this basis" (p. 64).

The results of careful analyses of the content of beginning reading programs, such as those of Beck and Willows et al., make it clear that there are important differences among programs and that these differences reflect different assumptions about *something*. It is possible that this is not a matter of different assumptions about the goals of reading instruction nor even about how reading works. Both code-emphasis programs and meaning-emphasis programs agree that comprehension is the objective of reading. Nevertheless, it is true that different aspects of the reading process are emphasized in order to justify different programs. For example, Goodman's (1967) description of reading as a "psycholinguistic guessing game" is seen by some basal proponents as support for basal programs. (Of course, the meaning-emphasis approach to reading instruction is much older than the "psycholinguistics" perspective.)

Goodman, in fact, argued that young children may read more meaningful, and hence more complex, material better than simplified controlled texts. The idea is that complex texts give more syntactic and semantic cues that (even allowing for many errors on individual words) will make a better match with the child's existing linguistic abilities. However, it does not appear that the call for more complex texts has been heeded, even by basal programs. Thus the most striking differences between basal programs and code programs seem not to reside in their discourse features, e.g., their syntax or their narrative structure. The differences are most noticeable in their vocabulary and their attention to correspondences.

The meaning-emphasis programs may also have drawn on theories of word perception. Williams (1979) points out that in the period roughly from 1920 to 1950, meaning-emphasis advocates pointed to evidence that a word is read "as a whole" without recognition of its constituent letters. Williams also suggests that the work of Gestalt psychologists in that period may have been influential. The Gestaltists emphasized holistic perceptual processes that are congruent with our somewhat misleading intuition that we perceive words as wholes. Frank Smith's word perception model (Smith, 1971), based on information processing concepts, also could be seen as consistent with the whole-word approach.

However, it seems fair to say that a general orientation to teaching and learning, rather than a careful reading of research, is the main source of meaning-emphasis approaches. There is an emphasis on meaning and comprehension and on the value of making text experiences as meaningful as possible. Research on reading processes certainly does challenge some of the assumptions of this approach, but its impact does not yet appear to be great. Meanwhile, the teaching of reading is a practical matter that involves complex decisions to use particular programs. These programs, in turn, usually reflect little influence of research, although there are some exceptions. As Willows et al. (1981) concluded from their content analysis of widely used programs, publishers' decisions about all important features of their programs "must be largely based on intuitions, arbitrary decisions, and marketing considerations" (p. 167).

The "psycholinguistic" approach

Some of the most popular justification for meaning-emphasis approaches comes from the work of Smith and Goodman. In 1973, Frank Smith's *Psycholinguistics and Reading* was published. Earlier, Kenneth Goodman had published a paper in the *Journal of the Reading Specialist* entitled "Reading: A Psycholinguistic Guessing Game" (1967). In 1971 Smith and Goodman together expounded "On the Psycholinguistic Method of Teaching Reading" in the *Elementary School Journal* (reprinted in Smith's book). Thus between 1967 and 1973 "psycholinguistics" had been applied by Goodman and Smith to reading processes, reading problems, and reading instruction.

The reading process, according to Goodman, is a "psycholinguistic guessing game" in which the reader figures out meanings. There are three cueing systems in this game. The graphonic, the syntactic, and the semantic (Goodman, 1970). The *graphonic* cues are not just what the reader knows about grapheme-phoneme correspondences, but rather general knowledge of spelling-sound relations. The *syntactic* cueing system is the reader's knowledge of syntactic patterns and the markers that cue these patterns, such as function words and inflectional suffixes. The *semantic* system is everything else—the reader's knowledge of word meanings, knowledge of the topic, everything. (No doubt if *pragmatics* had the currency in the 1960's that it has today, there would have been a fourth cueing system.) The "psycholinguistic" approach to reading was one that emphasized the reader's use of all relevant information in the attempt to get meaning. It is a sort of interactive model, but without specific suggestions as to how things interact.

The main failing of this approach to reading is that it does not recognize that one of the "cueing systems" is more central than the others. A child who learns the code has knowledge that can enable him to read no matter how the semantic, syntactic, and pragmatic cues might conspire against him. No matter how helpful they are to reading, these cues are not really a substitute for the ability to identify a word.

The instructional implications of the Goodman-Smith approach are not very specific. In fact, if there are "psycholinguistic teachers" and "psycholinguistic methods" in schools, these are not necessarily what Goodman and Smith wanted. They explicitly renounced "psycholinguistic instruction" in their 1971 paper (see Smith, 1973), which they called a "preemptive strike" against the misuse of their ideas. In fact, the only important instructional principle seems to be that phonics not be taught. It really does not matter much what is done in a classroom, just so long as certain things are not done.

The following words perhaps typified their published attitudes on this issue: "The value of a psycholinguistic approach" to reading would be the "very antithesis of a set of instructional materials." Instead, "the key factors of reading lie in the child and his interaction with information-providing adults. . . . Materials most compatible with such interaction are those that interfere least with natural language functioning" (Smith and Goodman, pp. 178 of Smith, 1973).

Overall, it is fair to say that the "psycholinguistic" approach is short on specifics. It seems compatible with any meaning-emphasis approach. However, it may be more accurate to say that it is incompatible with any approach that includes code instruction as a significant part of teaching.

Research on beginning reading instruction

Jeanne Chall (1967) tried to address the practical question of reading instruction. Is beginning instruction more successful with code-emphasis or meaning-emphasis programs?[1] The answer to "the great debate" favored the code-emphasis programs, based on Chall's analysis of hundreds of method comparison studies. Consistent with their designs, code-emphasis programs produced better performance in word recognition and spelling. Children in code-emphasis programs were not at a disadvantage even in their interest in reading, ostensibly one of the advantages of meaning emphasis.

There have been other projects to compare instructional methods. An advantage for a particular program is not always found (Bond & Dystra, 1967), and when an advantage is found it may be subject to other interpretation (House, Glass, McLean, & Walker, 1978). Furthermore, it is likely that the effectiveness of particular reading instruction depends on time spent in reading instruction and students' socioeconomic class (Guthrie, Martuza, & Seifert, 1979). Williams (1979), reviewing the evidence based on comparative studies, concluded that the measurable advantages of code-emphasis programs for reading achievement were small, confined to word recognition and spelling, and largely restricted to the first two grades. However, it is even more clear that there is no measurable advantage of meaning-emphasis programs even in measures of comprehension.

There is no getting around the conservative conclusion that code-emphasis programs do not hurt comprehension and provide at least some help for word recognition. Williams (1979) offers some interesting ideas as to why this fact has not had more impact on how reading is taught. The reasons may include theoretical attractiveness—comprehension is "sophisticated," decoding is "simple-minded"—teacher professionalism, and social philosophy. However, as far as the evidence is concerned, the great debate seems over, and Chall's (1967) conclusions hold.

There are some recent comparative studies that reinforce the conclusion concerning code emphasis. One is the Pittsburgh longitudinal study described in the previous chapter. This study followed children in a code-emphasis classroom as well as children in a basal curriculum. The code-emphasis program was the NRS (Beck & Mitroff, 1972) developed at the University of Pittsburgh (see Chapter 10). The comparative data are still being analyzed, although reports of some data for each curriculum group are reported in Lesgold and Curtis (1981) and Lesgold and Resnick (1982).

One clear conclusion is available, based on the oral reading performance of students in the two programs. Errors in oral reading are almost always appropriate for the context, even in the code-emphasis program (Lesgold & Curtis, 1981), at least by the middle of the first grade. Thus the fear that code-emphasis programs produce "word callers" who are insensitive to contextual meaning was not confirmed.

A second longitudinal study was reported by Evans and Carr (1983). They evaluated two programs in 20 first-grade classrooms. Half of these were traditional teacher-directed classrooms. Instruction used basal readers with "phonics" drills and applications. The other half of the classrooms were taught in less traditional student-centered classrooms, in which teacher-led instruction constituted only 35% of the day's activity. Reading was taught by an individualized language-experience method. Students produced their own books of stories and banks of words to be recognized. Evans and Carr (1983) characterize these two groups as decoding-oriented and language-oriented. Evans and Carr note that despite differences in how reading was taught, the two groups did not differ in the amount of time students spent on reading tasks. The two groups of classrooms were matched on relevant social variables and were virtually identical on measures of intelligence and language maturity. However, the clear result was that the decoding group scored higher on year-end reading achievement tests, including comprehension tests. Furthermore, the language-experience group did not show any higher achievement in oral language measures, based on a storytelling task.[2]

Thus comparative studies continue to support the assumption that instruction that includes code teaching is advantageous for some things (reading) and not harmful for others. There may be gains from language-experience (or meaning-emphasis) programs that escape the kinds of measures conventionally used. But this is a weak argument against the central point that ensuring that a child learns about the code is a worthwhile instructional objective.

COMPREHENSION

Part of reading instruction, especially beyond the beginning stages, is directed at comprehension. In fact, some of the meaning-emphasis programs tend to suggest that comprehension is being taught from the first lesson. On the other hand, it is not clear that reading lessons actually do much *teaching* of comprehension compared with monitoring and evaluation of student reading behaviors (Durkin, 1978–1979). In any case, from a theoretical perspective, there is a prior question: Can comprehension be taught?

For this question to be meaningful, a distinction must be made between knowledge and comprehension strategies. A parallel distinction is that between indirect teaching of comprehension and direct teaching of comprehension. If a student is constantly asked an inference question after reading a story, one effect may be to cause the student's routine comprehension processes to become inferential. If so, this would be indirect teaching of comprehension. Alternatively, the reading and hearing of numerous simple stories may lead a child to implicitly understand the structure of narratives. The child may be expected to begin to use the schema for a story to guide comprehension and memory for stories. In these cases comprehension processes are strengthened or modified by implicit induction of patterns available in language and in the event world that language maps. It is quite likely that this is the ordinary manner for learning comprehension.

Nevertheless, it is also quite likely that there are some obstacles to applying comprehension processes to written text, at least for some children. Preschool language is different in some important ways from literate language. Most important, among the many differences, is that a child's preschool spoken language is heavily *contextualized*.[3] That is, spoken language occurs always in a communication setting in which there are things to be referred to, a communication partner, and an intrinsically important message. The written texts that confront the student in school are quite different. Reference in the student's world is lacking, there is no communication partner with a shared social interest, and the message itself often has only extrinsic value to the student. Thus written language, in schools and elsewhere, is *decontextualized*.

One consequence of decontextualized language is that comprehension processes are almost completely dependent on the text. In a sense, the meaning is in the text and not in the event world. This means, among other things, that inference processes that in spoken communication might be triggered by some element of the context, must be triggered by a written text element. It is quite possible that normal language inferences and appropriate text knowledge are not readily activated for some young readers. This possibility is acknowledged by reading programs that expect the teacher to provide relevant background knowledge for stories read by students and by teachers who do such things on their own.

A second possibility, perhaps even more serious for comprehension, is that the syntactic structures encountered in print are not under the reader's control. One of the costly luxuries of contextualized language is that there are few demands placed on syntactic processing. Meanings are carried by context more than by syntax. The child entering school may have a basic syntactic knowledge but not necessarily one up to the demands written texts will make.

Such considerations argue against the assumption that young readers have all the comprehension tools necessary to take on written language. Still, the possibility of some comprehension problems does not imply that direct teaching of comprehension strategies is called for. It remains likely that for many students the comprehension processes will become more inferential, more syntactic, and generally more text-based without special instruction. This is especially likely to the extent that children become fluent at lexical access. As the decoding hurdle is overcome, it is a lot easier for a child to discover text inferences. A basic problem for a child with inefficient decoding is that he may not have enough information in memory at any one time to make needed inferences.

It is also important to realize that some problems in comprehension are due to lack of specific knowledge. The reader cannot understand a text without having control over most of the concepts in the text. This means that specific concept knowledge (vocabulary) and more elaborated general concept knowledge (schemata) are critical to comprehension. Again, vocabulary learning and relevant background knowledge can be provided by a teacher. And the acquisition of knowledge in and out of school is part of what will make a student good at reading comprehension.

If there are so many factors to consider in comprehension, the problem of teaching comprehension is serious. It is especially questionable to recommend specific comprehension instruction for children whose lack of decoding fluency is their major problem. As a practical matter, it would be better to be able to teach comprehension *and* to strengthen decoding skills rather than to have to choose between them. Beck (1981, 1983) has suggested a way of making the more agreeable choice. She proposed a two-strand reading instruction program, with one strand consisting of conceptually easy material and a second strand consisting of more difficult material. The easier material would encourage decoding fluency. The more difficult material, which would share a common domain of knowledge, would allow students to stretch their knowledge base and perhaps their higher order comprehension skills. Whatever the specifics, some recognition that word-level fluency must continue to increase for many children throughout the elementrary years is essential. The insight is actually quite simple: Comprehension depends on many things, and one of them is fluent word recognition.

Finally, consider again direct teaching of comprehension strategies. The points raised above emphasize the importance of decoding fluency and knowledge. On the other hand, another point was that some higher order comprehension skills might not be adequate for spontaneous uses with print. It is the latter that are candidates for direct comprehension instruction. Proposals for direct teaching of comprehension strategies abound, particularly the teaching of comprehension monitoring (Brown, 1978; Collins & Smith, 1982). Such instruction is probably important for many students. As a practical matter, however, it needs to be considered in the same context as the decoding vs. knowledge issue that Beck (1981, 1983) addressed. That is, the teaching of comprehension monitoring should not be a substitute for other reading activities, in the absence of evidence that monitoring of comprehension is a fundamental problem for the student. These issues emerge again in the next section.

IMPROVING READING OF LOW-ACHIEVING READERS

In this final section we consider this question: How can the reading skills of low-achieving readers be improved? The issues of decoding, specific knowledge, and comprehension strategies remain as they do in regular reading instruction. Verbal efficiency theory claims that decoding inefficiency will be a contributor to poor comprehension. Gains in decoding are needed for most low-achieving readers to improve. However, since other components of comprehension are likely to be less well developed in such readers, it is likely that gains in achievement can be brought about by gains in any component—to the limit of the child's verbal coding efficiency. First we consider teaching comprehension, then verbal coding.

Comprehension Instruction

The central assumption of reading comprehension instruction is that low-skill readers suffer from deficits in comprehension strategies (Brown & Palinscar, 1982; Golinkoff, 1976; Ryan, 1982). The evidence for this assumption is quantitatively substantial (Ryan, 1982), but the theoretical status of comprehensive deficits is uncertain. That is, we expect that less able readers fail to perform well on many tasks, including tasks sensitive to comprehension strategies. Generally, however, studies showing comprehension deficits have not taken decoding fluency into account. (A few studies have taken decoding accuracy into account.) It has yet to be demonstrated that there are individuals who have comprehension strategy deficits without decoding fluency problems.[4] The case for a fundamental role for comprehension strategy deficits will be strengthened if such readers are found. A second reason for caution regarding the strategy deficit assumption is that some studies that look for strategy deficits do not find them (Weaver & Dickinson, 1982; Townsend, 1982). It is never the case that one looks for decoding fluency problems in low-ability readers and fails to find them.

On the other hand, a better approach may be in instructional research.

Can low-achieving readers be taught comprehension strategies that will improve their reading comprehension? If such improvements can be found, it does not matter that some other problem remains. It will mean that gains can be made up to the limits imposed by other factors.

Comprehension-monitoring strategies

There have been some success stories in teaching comprehension monitoring (Meichenbaum & Asarnow, 1978; Palinscar & Brown, 1984). The Palincsar and Brown study taught seventh-grade low-achieving readers monitoring procedures to apply to texts. For example, one skill was summarizing, following the observation of Brown, Day, & Jones (1983) that younger readers do not have adequate summarizing skills. Questioning, clarifying, and predicting were other skills that children were taught. Although it is not possible to tell what was instrumental within the instructional program, it is clear that some gains were achieved. Not only did children learn to perform the monitoring tasks—for example, to produce better summaries—but the gains generalized to other measures of text comprehension, including standardized test scores.

The concept of comprehension monitoring, of course, covers a lot of territory. In a sense it is a fairly standard instructional device, as evidenced by many programs that encourage students to ask themselves questions, state main points, etc. Beyond such general activities, comprehension monitoring can be understood by reference to activities of text processing. Thus "summarization" is a more useful description of an activity than "critical thinking" or "remember the main points." Summarization requires not only obtaining gist from the text but also inferencing and organization. It may be the kind of activity that focuses attention to the text as meaning bearer.

What about inferential text processing? The processes of filling in text gaps are a large part of reading comprehension, and this is one of the things that some readers might not do spontaneously. Again there is evidence that low-achieving readers can be taught to improve these skills. In a series of studies, Bransford, Franks, Stein, and others examined some differences between high- and low-achieving fifth-grade students (see Bransford, Stein, Vye, Franks, Auble, Mezynski, & Perfetto, 1982). One of several interesting results concerns the allocation of study time.

Two reading passages were similar except that one was more explicit. The less explicit text omitted elaborations that made some relationships more difficult to see. For example, in a text describing a robot, the explicit version included these sentences: "It had a bucket on top of its head in order to carry paint. This robot was carrying a roll of tape to put on the windows to protect them from the paint." The implicit version included these sentences in place of those above: "It had a bucket on top of its head. The robot was carrying a roll of tape." The difference is that in the explicit version the purpose of the bucket and of the tape are clear. In the implicit version, these relations must be inferred.

The study time results were that high-achieving students spent more time studying the implicit passage. Low-achieving students spent more time studying the explicit passage, which was longer. Clearly spending more time on a text to fill in the inferential gaps is a good strategy to use, and low-achieving students did not do it.

Training the low-achieving students consisted of several procedures carried out with isolated arbitrary sentences. Students were prompted to evaluate the arbitrariness of sentences and to generate a context that would help them understand an arbitrary sentence. (An example of an arbitrary sentence is "The tall man bought the crackers.") Prompts were discontinued as soon as possible with the intention that the students begin to apply these elaboration procedures spontaneously. The results of this training were seen in students' study times for an implicit text. They spent more time studying the implicit text following training, and their cued recall of the text also improved. The conclusions are that some inference processes may not be triggered by texts for low-achieving readers and that training to use text elaboration procedures can be effective.[5]

Story structures

A final example of remedial comprehension training is story structures. Most of the reading material in the elementary grades comes from stories. It is possible to wonder whether low-achieving readers might be handicapped by an inadequate knowledge of the structure of stories. However, there is little reason to assume this is the case in general. The tacit knowledge that stories conform to general syntactic structure—a setting followed by an embedding episode structure—is probably universally acquired within a culture. We can expect some variance in the age of acquisition given current methods for assessing knowledge of story structures. However, by school age most children have an adequate implicit knowledge of narrative forms (Stein & Glenn, 1979), although the richness of the knowledge may continue to develop. In fact, it is unlikely that older low-achieving readers have a significant story knowledge problem (Weaver & Dickinson, 1982).

Nevertheless, there have been attempts to improve comprehension of low-achieving readers through story grammar training. An attempt to train fifth-graders by Dreher and Singer (1980) was unsuccessful, but at least two studies training fourth-graders have been successful. Fitzgerald and Spiegel (1983) found that low-achieving students benefited from direct instruction on story structure. The gains were seen both in their knowledge of story structures measured in production and in their comprehension of stories. Interestingly, effects were obtained rather quickly, within six 30-minute sessions over a two-week period, and they did not increase with more long-term instruction. A similar gain in comprehension of low-achieving students with very short-term instruction on story structures was reported by Short and Ryan (1982).

Although it is not certain which features of instruction are important here, a clear, if tentative, conceptualization of the problem may be possible. The low-achieving readers did not lack knowledge of story structures. In both studies, their gains were due to a relatively small dose of instruction that, at least temporarily, made their story knowledge more useful to them. Whether such gains can be expected to endure is unknown. A practical conclusion seems to be that some instruction in story structures might be helpful for low-achieving readers, but that it should be seen as a short-term intervention aimed at helping students apply their knowledge to written stories. There are no grounds for making story grammar instruction a central component of a remedial program.

Decoding Instruction

One implication of verbal efficiency theory is that improving a low-achieving reader's decoding should allow improvements in comprehension. However, as noted at the outset of this chapter, this does not mean that training on decoding speed can necessarily be expected to be advantageous. Its success depends on whether the student will acquire knowledge or procedures that serve rapid decoding or whether he merely learns to respond more quickly. Merely training a student to say words quickly will not necessarily result in better comprehension, as the research has confirmed (Piggins & Barron, 1982; Fleisher, Jenkins, & Pany, 1979).[6] On the other hand, practice at actual reading may strengthen both decoding and comprehension. Samuels (1979), for example, reports that children with difficulty in reading benefit from systematic rereading of meaningful passages. This "repeated readings" procedure raises the reader's fluency and maintains comprehension.

Instruction in decoding can be successful, providing certain obstacles are overcome. One obstacle is the potential of decoding tasks to be seen as dull routines of drill and practice. The negative impact on student *and teacher* motivation is an obstacle. As a matter of fact the motivational components of any program are probably profound. Gains attributed to comprehension instruction (e.g., story structures) may well derive in part from the engaging of both teachers and students in activities that increase motivation for instructional tasks. Decoding instruction will not realize this important advantage unless some thought is given to the problem.

A related obstacle to decoding instruction is that it requires long-term instruction. Instruction in comprehension strategies relies, in part, on "reminding"—that is, allowing the student to use knowledge that he may already possess in a less useful form. But fluent word processing depends on a base of knowledge that is not easily acquired outside of reading. It relies also on two separable levels of knowledge, knowledge of word forms and procedures for efficient access to these word forms. Roughly speaking, this is the accuracy and speed distinction in decoding. The procedures for rapid access have to apply to a well-developed verbal coding system that includes countless lexical entries, orthographic patterns, and associated phonemic patterns.

For the skilled reader this verbal knowledge base is built up gradually over reading experiences, regardless of how the initial reading experiences were organized. The skilled reader's lexical access processes become fluent interactive processes in which overlearned patterns with redundant links (e.g., letters and associated phonemes) play a central role (see Chapters 2 and 4). The low-achieving reader starts out behind in terms of some of the linguistic knowledge on which this verbal processing system gets built. He falls farther behind as his reading experiences fail to build the rich and redundant network that the high-achieving reader has. By the time a fifth-grade student is targeted for remediation, the inefficiency (and ineffectiveness) of his (or her) verbal coding system has had a significant history. To expect this to be remedied by a few lessons in decoding practice is like expecting a baseball player of mediocre talent to suddenly become a good hitter following a few days of batting practice. This problem, the need for extended practice, is unfortunately coupled with the problem of motivation. The component that needs more than a few days work risks tedium if carried out for more than a few days. Of course, this is a solvable problem where there is a will to solve it. A general solution is to embed word practice in story reading lessons during all stages of reading instruction, as Beck (1981) suggested.

In fact, there are many solutions to the problems of decoding instruction and many of them have been widely used in remedial practice. One solution involves computer-based practice. The motivational features of computer-based games are obvious and can be used to advantage. The record-keeping and feedback value of computer instruction are also widely acknowledged. (For examples of computer instruction in reading, see several chapters in Wilkinson, 1983, especially those of Lesgold and Wilkinson and Patterson.)

A decoding remediation project
One good example of computer-based decoding instruction comes from a training study by Frederiksen and others (Frederiksen, Weaver, War-

ren, Gillotte, Rosebery, Freeman, & Goodman, 1983). The subjects of the training study were 10 secondary school students who were very low in reading skill. Their reading scores on the Nelson Denny Reading Test ranged from the 4th through the 32nd percentile. The training tasks were designed to provide instruction on three components of reading that Frederiksen (1981) had identified in his analysis of reading skill. These components are (1) perception of multiletter units; (2) efficient phonological decoding of orthographic information; and (3) use of context frames in accessing and integrating meanings. The first two are basic lexical access processes that need to be automatically executed for fluent reading to occur. The third is the use of context to activate the appropriate meaning of a word.

The main objective of the training study was to increase the student's speed of processing in these reading components. The assumption is that such training can develop the student's capacity for automatic performance. The training procedures have several features worth noting. First, training takes place in a computer game environment. Each of the games has elements of clear game objectives, competition, and fantasy. Second, the training involved several thousand trials, or experiences with individual words. Third, students were not allowed to sacrifice accuracy for speed. Finally, attention was given to ordering the training tasks within a theoretically motivated hierarchy.

For illustration, consider the task to teach efficient decoding, a game called Racer. In Racer, the student faced a display in which a 20-word matrix appeared, one word at a time. The race is between a sailboat, controlled by the student's rate of decoding, and a horse, whose speed is set by the student's previous success in races. The race is seen on the screen as the horse and sailboat move toward a finish line. The student who wins runs against a faster horse in the next race. Inaccuracy, however, is penalized.

The materials for the Racer game increase in difficulty over a series of matrices. Within a race matrix, the words represent a mix of decoding rules (e.g., different vowels). Especially interesting, however, is the individual flexibility that seems to be required. Frederiksen et al. (1983) report that two subjects had difficulty with the standard materials. Accordingly, they were first given matrices that consistently used a single decoding rule and then moved to the standard matrices. This seems to reflect a point emphasized earlier: Speed training cannot be expected to work unless it is actually teaching underlying orthographic knowledge or unless such knowledge is already available to the student.

The results of training on the three game-environment tasks were impressive, although somewhat complex. There were only 10 subjects and only 4 who received all three training tasks: multiletter perceptual training (Speed), decoding (Racer), and context frames (Ski Jump). All subjects, regardless of which combination of training tasks they received, showed gains on criterion tests that measured what the tasks were designed to practice. For example, students who were trained on Racer showed gains on both a word-reading task and a pseudoword-reading task. These gains included both speed and accuracy, indicating that training affected basic knowledge relevant for decoding as well as speed. On the other hand, gains were not obtained in a task designed to test multiletter perception. This task required subjects to detect the presence of target letters in words presented for very brief (150 millisecond) exposures. Students who received training on perception of multiletter units did show gains on this task. Thus, training was specific in the sense that training on a higher level task (phonetic decoding) did not transfer to performance on a lower level task (multiletter decoding). Similarly, training on use of context frames, the highest level, did not transfer down to decoding unless the subjects were first trained on decoding. Frederiksen et al. (1983) conclude that after some gains have been achieved in a lower level skill *on a task that focuses just on that skill,* the student can continue to improve in a higher level task. But it must get a start in the lower level task first.

On the other hand, there was inconclusive evidence of transfer of training from the word processing tasks to comprehension. Of the four subjects who went through all training tasks, three showed some gains in reading rate measured by an inference task (one with low accuracy, however). The one subject who did not show a gain in rate showed an increase in the Nelson Denny reading test, which includes comprehension. The small number of subjects and the corresponding salience of individual differences clearly suggest a very cautious conclusion on the issue of transfer to comprehension. However, it is fair to conclude that the Frederiksen et al. project demonstrates dramatic gains in decoding ability and at least some transfer to higher level reading tasks.

The value of computer-based decoding training is being explored in other projects. In a project currently underway in Pittsburgh, decoding games are designed for children.[7] As in the Frederiksen et al. project, an objective is to increase the automaticity of decoding as well as the knowledge that decoding requires. And, also as in the Frederiksen et al. project, this goal is met in a computer game environment which sustains student interest. For example, in one game, the student builds words out of letters arranged in a matrix on a touch-sensitive screen. The child touches the initial letters on the left column of the matrix and the endings in the cells of the matrix that will make a word. For example, the child touches *b* from the left column and *ar* and *ut* from the matrix to make *bar* and *but*. The words the child produces are stacked up visually, and in one version of the game, the child tries to better the speed of his or her best previous game. There are other games designed around the general principle of providing speeded practice in a game setting, always with careful attention to sequencing materials. Although the effectiveness of the program cannot yet be evaluated objectively, it is likely that such a program, based on sound cognitive and motivational principles, will succeed in improving decoding skills of low-achieving readers.

It should go without saying that both the demonstrated and anticipated successes in decoding programs are not essentially dependent on computer instruction. The sound principles of decoding practice can be built into many formats.

SUMMARY

The teaching of reading is a complex subject not simply directed by knowledge of reading processes. However, the theory of verbal efficiency implies that objectives of reading instruction should include high levels of skill at lexical access, including decoding, that are more or less automatic. Mere training of speeded word naming is not implied, because the underlying knowledge that allows fluent lexical access is often the critical problem.

For beginning reading, it seems clear that decoding instruction should be a primary objective. There is no known negative result from programs that emphasize decoding and there are at least some positive results in the early grades.

Comprehension instruction may also be beneficial, and it is a mistake to conclude that reading instruction should ignore either decoding or comprehension. It is possible that some comprehension abilities have not been adequately developed in the preschool years or that difficulties in handling written text prevent these abilities from being used. For example, it appears likely that instruction in strategies for making text inferences, for summarizing texts, and for noticing structural elements may be useful. There is, however, no evidence that such procedures fully compensate for inefficient word processing. They are probably effective only within the limits placed by a reader's word processing.

Decoding training can be successful even for students beyond beginning reading. What is needed for success is careful attention to the theoretical underpinnings of a training task along with procedures to sustain student and teacher interest. Questions of transfer to comprehension still remain, but it is clear that decoding training can be very successful at making word processes more efficient and at least partially successful in improving other components of reading achievement.

NOTES

1. The distinction between code emphasis and meaning emphasis should not be too broadly drawn, according to Chall (1967). The difference is mainly one of when letter-sound correspondences are introduced and with what emphasis. Venezky (1978) made a similar point. Thus, the differences may be considered technical rather than fundamental. But "technical" should not be equated with "trivial."

2. According to the authors of this study, the educational district that funded the study wanted an outcome other than the one obtained. The report has been ignored, if not actually suppressed by its sponsors. This is, one hopes, an atypical example of the difficulty encountered in having research make a positive impact in education.

3. This point is made in a bit more detail in Perfetti (in press). For other discussions of the differences between spoken and written language, see Olson (1977), Rubin (1980), and Tannen (1982).

4. This claim is made with appreciation of Steiner, Wiener, and Cromer (1971) which is often cited as evidence that decoding and comprehension are disassociated (also Cromer, 1970).

5. Unfortunately the training study did not compare post-training performance on implicit texts vs. explicit texts. Thus we cannot really conclude that training made students more sensitive to features of text.

6. Tom Hogaboam and I did an unpublished training study that showed that low-skill third-graders could raise their word-naming speed after two days of practice. Following the practice they did not answer comprehension questions for passages containing the taught words any better than control subjects. It is necessary to train lexical knowledge or lexical access facility rather than speed per se.

7. This project has been developed by Isabel Beck, Steve Roth, Robert Glaser, and Alan Lesgold.

References

Aaron, P. G. (1978). Dyslexia, an imbalance in cerebral information processing strategies. *Perceptual and Motor Skills. 48*, 699–706.

Allport, D. A. (1977). On knowing the meaning of words we are unable to report: The effects of visual masking. In S. Dornic (Ed.), *Attention and performance VI* (pp. 505–533). Hillsdale, NJ: Erlbaum.

Anderson, J. R. (1976). *Language, memory, and thought*. Hillsdale, NJ: Erlbaum.

Anderson, R. C., & Freebody, P. (1979). *Vocabulary knowledge* (Technical Report No. 146). Urbana: University of Illinois Center for the Study of Reading.

Anderson, R. C., Reynolds, R. E., Schallert, D. L., & Goetz, E. T. (1976). *Frameworks for comprehending discourse* (Technical Report No. 12). Urbana: University of Illinois Laboratory for Cognitive Studies in Education.

Anderson, R. C., Spiro, R. J., & Anderson, M. C. (1978). Schemata as scaffolding for the representation of information in connected discourse. *American Educational Research Journal, 15*, 433–440.

Atkinson, R. C., & Shiffrin, R. M. (1968). Human memory: A proposed system and its control processes. In K. W. Spence & J. T. Spence (Eds.), *The psychology of learning and motivation* (Vol. 2). New York: Academic Press.

Baddeley, A. D., & Hitch, G. T. (1974). Working memory. In G. A. Bower (Ed.), *The psychology of learning and motivation* (Vol. 8). New York: Academic Press.

Baddeley, A. D., & Lewis, V. J. (1981). Inner active processes in reading: The inner voice, inner ear, and inner eye. In A. M. Lesgold & C. A. Perfetti (Eds.), *Interactive processes in reading*. Hillsdale, NJ: Erlbaum.

Bakker, D. J. (1979). Hemispheric differences and reading strategies: Two dyslexias? *Bulletin of the Orton Society, 29*, 84–100.

Baron J. (1977). Mechanisms for pronouncing printed words: Use and acquisition. In D. LaBerge, & S. J. Samuels (Eds.), *Basic processes in reading: Perception and comprehension* (pp. 175–216). Hillsdale, NJ: Erlbaum.

Baron, J. (1978). The word-superiority effect: Perceptual learning from reading. In W. K. Estes (Ed.), *Handbook of Learning and Cognitive Processes, Vol. 6* (131–166). Hillsdale, NJ: Erlbaum.

Baron, J. (1979). Orthographic and word specific mechanisms in children's reading of words. *Child Development, 50*, 60–72.

Barron, R., & Baron, J. (1977). How children get meaning from printed words. *Child Development, 48*, 587–594.

Barron, R. W. (1981). Reading skill and reading strategies. In A. M. Lesgold &

C. A. Perfetti (Eds.), *Interactive processes in reading*. Hillsdale, NJ: Erlbaum.

Bartlett, F. C. (1932). *Remembering: A study of experimental and social psychology*. Cambridge University Press.

Bauer, D., & Stanovich, K. E. (1980). Lexical access and the spelling-to-sound regularity effect. *Memory and Cognition, 8*, 424–432.

Beck, I. L. (1981). Reading problems and instructional practices. In T. G. Waller & G. E. MacKinnon (Eds.), *Reading research: Advances in theory and practice* (Vol. 2). New York: Academic Press.

Beck, I. L. (1983). Developing comprehension: The impact of the directed reading lesson. In R. Anderson, R. Tierney, & J. Osborn (Eds.), *Learning to read in American schools*. Hillsdale, NJ: Erlbaum.

Beck, I. L. & Mitroff, D. (1972). *The rationale and design of a primary grades reading system for an individualized classroom*. University of Pittsburgh, Learning Research and Development Center Publication Series (No. 4).

Beck, I. L., Omanson, R. C., & McKeown, M. G. (1983). An instructional redesign of reading lessons: Effects on comprehension. *Reading Research Quarterly, 17*, 462–481.

Beck, I. L., Perfetti, C. A., & McKeown, M. G. (1982). Effects of long-term vocabulary instruction on lexical access and reading comprehension. *Journal of Educational Psychology, 74*, 506–521.

Becker, C. A. (1976). Allocation of attention during visual word recognition. *Journal of Experimental Psychology: Human Perception and Performance, 2*, 556–566.

Becker, C. A. (1980). Semantic context effects in word recognition: An analysis of semantic strategies. *Memory and Cognition, 8*, 493–512.

Belmont, L., & Birch, H. G. (1966). The intellectual profile of retarded readers. *Perceptual and Motor Skills, 22*, 787–816.

Bender, L. (1956). *Psychopathology of children with organic brain disorders*. Springfield, IL: Charles C. Thomas.

Benson, D. F. (1981). Alexia and neuroanatomical basis of reading. In J. Pirozzolo & M. C. Wittrock (Eds.), *Neuropsychological and Cognitive Processes in Reading*. New York: Academic Press.

Benton, A. (1975). Developmental dyslexia: Neurological aspects. In W. J. Friedlander (Ed.), *Advances in neurology 7* (pp. 1–47). New York: Raven Press.

Berger, N. S., & Perfetti, C. A. (1977). Reading skill and memory for spoken and written discourse. *Journal of Reading Behavior, 9*, 7–16.

Beverly, S. E., & Perfetti, C. A. (1982). *Phonemic and semantic processes of skilled and less skilled beginning readers*. Paper presented at the annual meeting of the American Educational Research Association, New York, NY.

Beverly, S. E., & Perfetti, C. A. (1983). *Skill differences in phonological representation and development of orthographic knowledge*. Paper presented at the Biennial Meeting of the Society for Research in Child Development, Detroit, MI.

Biemiller, A. (1970). The development of the use of graphic and contextual information as children learn to read. *Reading Research Quarterly, 6*, 75–96.

Birch, H. G. (1962). Dyslexia and maturation of visual function. In J. Money (Ed.), *Reading disability: Progress and research needs in dyslexia*. Baltimore, MD: Johns Hopkins University Press.

Birch, H. G., & Belmont, L. (1965). Auditory-visual integrative intelligence and reading ability in school children. *Perceptual Motor Skills, 20*, 295–305.

Boder, E. (1971). Developmental dyslexia: A dignositic screening procedure based on three characteristic patterns of reading and spelling. In B. Bateman (Ed.), *Learning Disorders* (Vol. 4). Seattle: Special Child Publications.

Boder, E. (1973). Developmental dyslexia: A diagnostic approach based on three atypical reading-spelling patterns. *Developmental Medicine and Child Neurology, 15*, 663–687.

Boder, E., & Jarrico, S. (1982). *The Boder reading-spelling pattern test: A diagnostic screening test for developmental dyslexia.* New York: Grune & Stratton.

Bond, G. L., & Dystra, R. (1967). The cooperative research program in first-grade reading instruction. *Reading Research Quarterly, 2,* 5–142.

Bouma, H., & Legein, C. P. (1977). Foveal and parafoveal recognition of letters and words by dyslexics and by average readers. *Neuropsychologia, 15,* 69–80.

Bouma, H., & Legein, C. P. (1980). Dyslexia: A specific recoding deficit? An analysis of response latencies for letters and words in dyslexics and in average readers. *Neuropsychologia, 18,* 285–298.

Bower, G. H., Black, J. B., & Turner, T. J. (1979). Scripts in memory for text. *Cognitive Psychology, 11,* 177–220.

Bradley, D. (1978). *Computational distinctions of vocabulary type.* Unpublished doctoral dissertation, Massachusetts Institute of Technology.

Bradley, L., & Bryant, P. E. (1983). Categorizing sounds and learning to read—a causal connection. *Nature, 301,* 419–421.

Brady, S., Shankweiler, D., & Mann, V. (1983). Speech perception and memory coding in relation to reading ability. *Journal of Experimental Child Psychology, 35,* 345–367.

Bransford, J. D., & Johnson, M. K. (1973). Considerations of some problems of comprehension. In W. G. Chase (Ed.), *Visual information processing.* New York: Academic Press.

Bransford, J. D., Stein, B. S., Vye, N. J., Franks, J. J., Auble, P. M., Mezynski, K. J., & Perfetto, B. A. (1982). Differences in approaches to learning: An overview. *Journal of Experimental Psychology: General, 111,* 390–398.

Breitmeyer, B. G., & Ganz, L. (1976). Implications of sustained and transient channels for theories of visual pattern masking, saccadic suppression, and information processing. *Psychological Review, 83,* 1–36.

Brewer, W. F., & Hay, A. E. (1982). *Reconstructive recall of linguistic style.* Unpublished manuscript, University of Illinois.

Britton, B. K., Meyer, B.J.F., Hodge, M. H., & Glynn, S. M. (1979). Effects of organization of text on memory: Tests of retrieval and response criterion hypotheses. *Journal of Experimental Psychology: Human Learning and Memory, 6,* 620–629.

Britton, B. K., Meyer, B. J. F., Simpson, R., Holdredge, T. S., & Curry, C. (1979). Effects of the organization of text on memory: Tests of two implications of a selective attention hypothesis. *Journal of Experimental Psychology: Human Learning and Memory, 5,* 496–506.

Brown, A. L. (1978). Knowing when, where, and how to remember: A problem of metacognition. In R. Glaser (Ed.), *Advances in instructional psychology* (pp. 77–165). Hillsdale, NJ: Erlbaum.

Brown, A. L. (1980). Metacognitive development and reading. In R. J. Spiro, B. C. Bruce, & W. F. Brewer (Eds.), *Theoretical issues in reading comprehension.* Hillsdale, NJ: Erlbaum.

Brown, A. L., Campione, J. C., & Barclay, C. R. (1979). Training self-checking routines for estimating test readiness: Generalization from list learning to prose recall. *Child Development, 50,* 501–512.

Brown, A. L., Day, J. D., & Jones, L. S. (1983). The development of plans for summarizing texts. *Child Development, 54,* 968–979.

Brown, A. L., & Palincsar, A. S. (1982). Inducing strategic learning from text by means of informed, self-control training. *Topics in Learning and Learning Disabilities, 2*(1), 1–17.

Brown, B., Haegerstrom-Portnoy, G., Adams, A. J., Yingling, C. D., Galin, D., Herron, J., & Marcus, M. (1983). Predictive eye movements do not discriminate between dyslexic and normal children. *Neuropsychologia, 21,* 121–128.

Bruce, D. J. (1964). The analysis of word sounds by young children. *British Journal of Educational Psychology, 34*, 158–170.

Butler, B., & Haines, S. (1979). Individual differences in word recognition latency. *Memory and Cognition, 7*, 68–76.

Byrne, B., & Shea, P. (1979). Semantic and phonetic memory codes in beginning readers. *Memory and Cognition, 7*, 333–338.

Calfee, R. C., Venezky, R. L., & Chapman, R. S. (1969). *Pronunciation of synthetic words with predictable and unpredictable letter-sound correspondences* (Technical Report No. 71). Madison: Wisconsin Research and Development Center for Cognitive Learning.

Caplan, B., & Kinsbourne, M. (1981). Cerebral lateralization, preferred cognitive mode, and reading ability in normal children. *Brain & Language, 14*, 349–370.

Carpenter, P. A., & Just, M. A. (1975). Sentence comprehension: A psycholinguistic processing model of verification. *Psychological Review, 82*, 45–73.

Carpenter, P. A., & Just, M. A. (1981). Cognitive processes in reading: Models based on readers' eye fixations. In A. M. Lesgold & C. A. Perfetti (Eds.), *Interactive processes in reading* (pp. 177–213). Hillsdale, NJ: Erlbaum.

Carroll, J. B. (1980). Individual difference relations in psychometric and experimental cognitive tasks. Chapel Hill: University of North Carolina, L. L. Thurstone Psychometric Laboratory, Rep. No. 163.

Chafe, W. L. (1982). Integration and involvement in speaking, writing, and oral literature. *Spoken and written language.* Norwood, NJ: Ablex.

Chall, J. (1967). *Learning to read: The great debate.* New York: McGraw-Hill.

Chase W. G., & Simon, H. A. (1973). Perception in chess. *Cognitive Psychology, 4*, 55–81.

Chiesi, H. L., Spilich, G. J., & Voss, J. F. (1979). Acquisition of domain-related information in relation to high and low domain knowledge. *Journal of Verbal Learning and Verbal Behavior, 18*, 257–274.

Chomsky, C. (1971). Write first, read later. *Childhood Education, 47*, 296–299.

Chomsky, C. (1977). Approaching reading through invented spelling. In L. B. Resnick & P. A. Weaver (Eds.), *Theory and practice of early reading, Vol. 2.* Hillsdale, NJ: Erlbaum.

Chomsky, N., & Halle, M. (1968). *The sound pattern of English.* New York: Harper & Row.

Cirilo, R. K., & Foss, D. J. (1980). Text structure and reading time for sentences. *Journal of Verbal Learning and Verbal Behavior, 19*, 96–109.

Clark, H. H., & Chase, W. G. (1972). On the process of comparing sentences against pictures. *Cognitive Psychology, 3*, 472–517.

Collins, A., & Loftus, E. (1975). A spreading-activation theory of semantic processing. *Psychological Review, 82*, 407–428.

Collins, A., & Smith, E. E. (1982). Teaching the process of reading comprehension. In D. K. Detterman & R. J. Sternberg (Eds.), *How and how much can intelligence be increased?* Norwood, NJ: Ablex.

Coltheart, M. (1978) Lexical access in simple reading tasks. In G. Underwood (Ed.), *Strategies of information processing* (pp. 151–216). London: Academic Press.

Coltheart, M. (1980). *Analyzing acquired disorders of reading.* Unpublished manuscript, Birkbeck College, University of London.

Coltheart, M., Besner, D., Jonasson, J. T., & Davelaar, E. (1979). Phonological recoding in the lexical decision task. *Quarterly Journal of Experimental Psychology, 31*, 489–507.

Coltheart, M., Davelaar, E., Jonasson, J. T., & Besner, D. (1977). Access to the internal lexicon. In S. Dornic (Ed.), *Attention and performance IV* (pp. 535–555). Hillsdale, NJ: Erlbaum.

Conrad, C. (1974). Context effects in sentence comprehension: A study of the subjective lexicon. *Memory and Cognition, 2*, 130–138.

Conrad, R. (1964). Acoustic confusions in immediate memory. *British Journal of Psychology, 55,* 75–84.

Critchley, M. (1964). *Developmental dyslexia.* London: Heinemann.

Cromer, W. (1970). A new explanation of some reading difficulties. *Journal of Educational Psychology, 61,* 471–483.

Cronbach, J. J. (1942). An analysis of techniques for diagnostic vocabulary testing. *Journal of Educational Psychology, 36,* 206–217.

Curtis, M. E. (1980). Development of components of reading skill. *Journal of Educational Psychology, 72,* 656–669.

Curtis, M. E. (1981). *Word knowledge and verbal aptitudes.* Unpublished manuscript, University of Pittsburgh, Learning Research and Development Center.

Curtis, M. E., & Glaser, R. (1983). Reading theory and the assessment of reading achievement. *Journal of Educational Measurement, 20,* 133–147.

Daneman, M., & Carpenter, P. A. (1980). Individual differences in working memory and reading. *Journal of Verbal Learning and Verbal Behavior, 19,* 450–466.

Davelaar, E., Coltheart, M., Besner, D., & Jonasson, J. T. (1978). Phonological recoding and lexical access. *Memory and Cognition, 6,* 391–402.

Davidson, B. J., Olson, R. K., & Kliegl, R. (1982, November). *Individual differences in developmental reading disability.* Paper presented at the meeting of the Psychonomic Society, Minneapolis, MN.

DeGroot, A. M. B. (1983). The range of automatic spreading activation in word priming. *Journal of Verbal Learning and Verbal Behavior, 22,* 417–436.

DeGroot, A. M. B., Thomassen, A. J. W. M., & Hudson, P. T. W. (1982). Associative facilitation of word recognition as measured from a neutral prime. *Memory and Cognition, 10,* 358–370.

Denckla, M. B., & Rudel, R. G. (1976a). Rapid "automatized" naming (R. A. N.): Dyslexia differentiated from other learning disabilities. *Neuropsychologia, 14,* 471–480.

Denckla, M. B., & Rudel, R. G. (1976b). Naming of object drawings by dyslexic and other learning disabled children. *Brain & Language, 3,* 1–15.

Dixon, P., & Rothkopf, E. Z. (1979). Word repetition, lexical access, and the process of searching words and sentences. *Journal of Verbal Learning and Verbal Behavior, 88,* 629–644.

Doehring, D. G. (1976). Acquisition of rapid reading responses. *Monographs of the Society for Research in Child Development, 41* (No. 165).

Dooling, D. J., & Lachman, R. (1971). Effects of comprehension on retention of prose. *Journal of Experimental Psychology, 88,* 216–222.

Downing, J. A. (1970). Children's concept of language in learning to read. *Educational Research, 12,* 106–112.

Downing, J. A. (1964). The I.T.A. (Initial Teaching Alphabet) reading experiment. *Reading Teacher, 18,* 105–109.

Downing, J. A., & Oliver, P. (1973–1974). The child's conception of "a word." *Reading Research Quarterly, 9,* 568–582.

Dreher, M. J., & Singer, H. (1980). Story grammar instruction unnecessary for intermediate grade students. *The Reading Teacher, 34,* 261–268.

Duane, D. D. (1979). Toward a definition of dyslexia: A summary of views. *Bulletin of the Orton Society, 29,* 56–64.

Durkin, D. (1966). The achievement of pre-school readers: Two longitudinal studies. *Reading Research Quarterly, 1(4),* 5–36.

Durkin, D. (1978–1979). What classroom observations reveal about reading comprehension instruction. *Reading Research Quarterly, 14,* 481–533.

Ehri, L. C. (1975). Word consciousness in readers and prereaders. *Journal of Educational Psychology, 67,* 204–212.

Ehri, L. C. (1976). Word learning in beginning readers and prereaders: Effect of form class and defining contexts. *Journal of Educational Psychology, 68,* 832–842.

Ehri, L. C. (1978). Beginning reading from a psycholinguistic perspective: Amalgamation of word identities. In F. B. Murray (Ed.), *The development of the reading process. International Reading Association Monograph* (No. 3). Newark, DE: International Reading Association.

Ehri, L. C. (1979). Linguistic insight: Threshold of reading acquisition. In T. Waller & G. E. MacKinnon (Eds.), *Reading research: Advances in theory and practice.* New York: Academic Press.

Ehri, L. C. (1980). The development of orthographic images. In U. Frith (Ed.), *Cognitive processes in spelling.* London: Academic Press.

Ehri, L. C. (1983a). A critique of five studies related to letter-name knowledge and learning to read. In L. M. Gentile, M. L. Kamil, & J. S. Blanchard (Eds.), *Reading Research Revisited.* Columbus, OH: Merrill.

Ehri, L. C. (1983b). Influence of orthography on phonological and lexical awareness in beginning readers. In J. Downing & R. Valtin (Eds.), *Language awareness and learning to read.* New York: Springer-Verlag.

Ehrlich, S. F., & Rayner, K. (1981). Contextual effects on word perception and eye movements during reading. *Journal of Verbal Learning and Verbal Behavior, 20,* 641–655.

Ellis, N. C., & Miles, T. R. (1978). Visual information processing in dyslexic children. In M. M. Gruneberg, R. N. Sykes, & P. E. Morris (eds.), *Practical aspects of memory.* London: Academic Press.

Evans, M. A., & Carr, T. H. (1983). *Curricular emphasis and reading development: Focus on language or focus on script.* Symposium conducted at the Biennial meeting of the Society for Research on Child Development. Detroit, MI.

Farnham-Diggory, S., & Gregg, L. W. (1975). Short term memory function in young readers. *Journal of Experimental Child Psychology, 19,* 279–298.

Fillmore, C. J. (1968). The case for case. In E. Bach & R. T. Harms (Eds.), *Universals in linguistic theory* (pp. 1–88). New York: Holt, Rinehart and Winston.

Fischler, I., & Bloom, P. (1979). Automatic and attentional processes in the effects of sentence context on word recognition. *Journal of Verbal Learning and Verbal Behavior, 18,* 1–20.

Fitzgerald, J., & Spiegel, D. L. (1983). Enhancing children's reading comprehension through instruction in narrative structure. *Journal of Reading Behavior, 14,* 1–18.

Flavell, J. H. (1976). Metacognitive aspects of problem solving. In L. B. Resnick (Ed.), *The nature of intelligence.* Hillsdale, NJ: Erlbaum.

Fleischer, L. S., Jenkins, J. R., & Pany, D. (1979). Effects on poor readers' comprehension of training in rapid decoding. *Reading Research Quarterly, 14,* 30–48.

Fletcher, J. M., & Satz, P. (1980). Developmental changes in the neuropsychological correlates of reading achievement: A six year longitudinal follow-up. *Journal of Clinical Neuropsychology, 2*(1), 23–37.

Fletcher, J. M., Satz, P., & Scholes, R. J. (1981). Developmental changes in the linguistic performance correlates of reading achievement. *Brain & language, 13,* 78–90.

Fowler, C. A., Wolford, G., Slade, R., & Tassinary, L. (1981). Lexical access with and without awareness. *Journal of Experimental Psychology: General, 110,* 341–362.

Fox, B., Routh, D. K. (1976). Phonemic analysis and synthesis as word-attack skills. *Journal of Educational Psychology, 68,* 70–74.

Frazier, L., & Fodor, J. D. (1978). The sausage machine: A new two-step parsing model. *Cognition, 6,* 291–325.

Frazier, L., & Rayner, K. (1982). Making and correcting errors during sentence comprehension: Eye movements in the analysis of structurally ambiguous sentences. *Cognitive Psychology, 14,* 178–210.

Frederiksen, J. R. (1978a). *Assessment of perceptual, decoding, and lexical skills and their relation to reading proficiency* (Technical Report No. 1). Cambridge, MA: Bolt Beranek and Newman.

Frederiksen, J. R. (1978b). *A chronometric study of component skills in reading* (Technical Report No. 2). Cambridge, MA: Bolt Beranek and Newman.

Frederiksen, J. R. (1981). Sources of process interactions in reading. In A. M. Lesgold & C. A. Perfetti (Eds.), *Interactive processes in reading.* Hillsdale, NJ: Erlbaum.

Frederiksen, J. R., & Kroll, J. F. (1976). Spelling and sound: Approaches to the internal lexicon. *Journal of Experimental Psychology: Human Perception and Performance, 2,* 361–379.

Frederiksen, J. R., Weaver, P. A., Warren, B. M., Gillotte, H. P., Rosebery, A. S., Freeman, B., & Goodman, L. (1983). *A componential approach to training reading skills* (Report No. 5295). Cambridge, MA: Bolt Beranek and Newman.

Fried, I., Tanguay, P. E., Boder, E., Doubleday, C., & Greensite, M. (1981). Developmental dyslexia: Electrophysiological evidence of clinical subgroups. *Brain & Language, 12,* 14–22.

Fry, M. A., Johnson, C. S., & Muehl, S. (1970). Oral language production in relation to reading achievement among select second graders. In D. J. Bakker & P. Satz (eds.), *Specific reading disability: Advances in theory and method.* Rotterdam University Press.

Gazzaniga, M. S. (1970). *The Bisected Brain,* New York: Appleton-Century-Crofts.

Gelb, I. J. (1952). *A study of writing.* University of Chicago Press.

Geschwind, N. (1979). Asymmetries of the brain—new developments. *Bulletin of the Orton Society, 29,* 67–73.

Gibson, E. J. (1969). *Principles of perceptual learning and development.* New York: Appleton-Century-Crofts.

Gibson, E. J., & Levin, H. (1975). *The psychology of reading.* Cambridge, MA: MIT Press.

Gleitman, L. R., & Rozin, P. (1973). Teaching reading by use of a syllabary. *Reading Research Quarterly, 8,* 447–483.

Gleitman, L. R., & Rozin, P. (1977). The structure and acquisition of reading. I: Relations between orthographies and the structure of language. In A. S. Reber & D. L. Scarborough (Eds.), *Toward a psychology of reading: The proceedings of the CUNY conference.* New York: Wiley.

Glowalla, U., & Colonius, H. (1983). Toward a model of macrostructure search. In A. Flammer & W. Kintsch (Eds.), *Discourse processing.* Amsterdam: North Holland.

Glushko, R. J. (1981). Principles for pronouncing print: The psychology of phonography. In A. M. Lesgold & C. A. Perfetti (Eds.) *Interactive processes in reading.* Hillsdale, NJ: Erlbaum.

Godfrey, J. J., Syrdal-Lasky, A. K., Millay, K. K., & Knox, C. M. (1981). Performance of dyslexic children on speech perception tests. *Journal of Experimental Child Psychology, 32,* 401–424.

Goldberg, R. A., Schwartz, S., & Steward, M. (1977). Individual differences in cognitive processes. *Journal of Educational Psychology, 69,* 9–14.

Goldman, S. R., Hogaboam, T. W., Bell, L. C., & Perfetti, C. A. (1980). Short-term retention of discourse during reading. *Journal of Educational Psychology, 68,* 680–688.

Golinkoff, R. M. (1976). A comparison of reading comprehension processes in good and poor comprehenders. *Reading Research Quarterly, 11,* 623–659.

Goodman, K. S. (1967). Reading: A psycholinguistic guessing game. *Journal of the Reading Specialist, 6,* 126–135.

Goodman, K. S. (1970). Reading: A psycholinguistic guessing game. In H. Singer & R. B. Ruddell (Eds.), *Theoretical models and processes of reading*, Newark, DE: International Reading Association.

Gordon, H. W. (1980). Cognitive asymmetry in dyslexic families. *Neuropsychologia, 18*, 645–656.

Gordon, H. W. (1983). Dyslexia. In R. E. Tarter & G. Goldstein (Eds.), *Neuropsychology of childhood*. New York: Plenum.

Gough, P. B. (1972). One second of reading. In J. F. Kavanaugh & I. G. Mattingly (Eds.), *Language by ear and eye: The relationship between speech and reading* (pp. 331–358). Cambridge, MA: MIT Press.

Gough, P. B., & Hillinger, M. L. (1980). Learning to read: An unnatural act. *Bulletin of the Orton Society, 20*, 179–196.

Graesser, A. C. (1981). *Prose comprehension beyond the word*. New York: Springer-Verlag.

Graesser, A. C., Gordon, S. E., & Sawyer, J. D. (1979). Recognition for typical and atypical actions in scripted activities: Tests of a script pointer + tag hypothesis. *Journal of Verbal Learning and Verbal Behavior, 18*, 319–333.

Graesser, A. C., Hoffman, N. L., & Clark, L. F. (1980). Structural components of reading time. *Journal of Verbal Learning and Verbal Behavior, 19*, 135–151.

Grice, H. P. (1975). Logic and conversation. In P. Cole & J. L. Morgan (Eds.), *Syntax and semantics, Vol. 3: Speech acts*. New York: Academic Press.

Guthrie, J. T., Goldberg, H. K., & Finucci, J. (1972). Independence of abilities in disabled readers. *Journal of Reading Behavior, 4*, 129–138.

Guthrie, J. T., Martuza, V., and Seifert, M. (1979). Impacts of instructional time in reading. In L. B. Resnick & P. A. Weaver (Eds.), *Theory and Practice of Early Reading, Vol. 3*. Hillsdale, NJ: Erlbaum.

Hammond, K. (1984). *Auditory and visual memory access and decoding in college readers*. Paper presented at the annual meeting of the American Educational Research Association, New Orleans, LA.

Hardyck, D. C., & Petrinovich, L. F. (1970). Subvocal speech and comprehension level as a function of the difficulty level of reading material. *Journal of Verbal Learning and Verbal Behavior, 9*, 647–652.

Haviland, S. E., & Clark, H. H. (1974). What's new? Acquiring new information as a process in comprehension. *Journal of Verbal Learning and Verbal Behavior, 13*, 512–521.

Helfgott, J. (1976). Phonemic segmentation and blending skills of kindergarten children: Implications for beginning reading acquisition. *Contemporary Educational Psychology, 1*, 157–169.

Heller, J. I., & Greeno, J. G. (1978, May). *Semantic processes in arithmetic word problem solving*. Paper presented at the Midwestern Psychological Association meeting, Chicago, IL.

Hildyard, A., & Olson, D. R. (1982). On the comprehension and memory of oral vs. written discourse. In D. Tannen (Ed.), *Spoken and written language: Exploring orality and literacy*. Norwood, NJ: Ablex.

Hillinger, M. L. (1980). Priming effects with phonemically similar words: The encoding-bias hypothesis reconsidered. *Memory and Cognition, 8*, 115–123.

Hogaboam, T. W., & Perfetti, C. A. (1978). Reading skill and the role of verbal experience in decoding. *Journal of Educational Psychology, 70*, 717–729.

Hogaboam, T. W., & Pellegrino, J. W. (1978). Hunting for individual differences in cognitive processes: Verbal ability and semantic processing of pictures and words. *Memory and Cognition, 6*, 189–193.

Hogaboam, T. W., & Perfetti, C. A. (1975). Lexical ambiguity and sentence comprehension. *Journal of Verbal Learning and Verbal Behavior, 14*, 265–274.

Holland, J. G. (1979). Analysis of behavior in reading instruction. In L. B. Resnick & P. A. Weaver, *Theory and practice of early reading, Vol. 1.* Hillsdale, NJ: Erlbaum.

Hood, J. (1976). Qualitative analysis of oral reading errors: The inter-judge reliability of scores. *Reading Research Quarterly, 11,* 577–598.

House, E. R., Glass, G. V., McLean, L. D., & Walker, D. F. (1978). No simple answer: Critique of the Follow Through evaluation. *Harvard Educational Review, 48,* 128–160.

Huey, E. B. (1908, 1968). *The psychology and pedagogy of reading.* Cambridge, MA: MIT Press, 1968. (Originally published 1908).

Hunt, E. (1978). Mechanics of verbal ability. *Psychological Review, 85,* 109–130.

Hunt, E., Davidson, J., & Lansman, M. (1981). Individual differences in long-term memory access. *Memory and Cognition, 9,* 599–608.

Hunt, E., Frost, N., & Lunneborg, C. (1973). Individual differences in cognition: A new approach to intelligence. In G. H. Bower (Ed.), *The psychology of learning and motivation* (Vol. 7). New York: Academic Press.

Hunt, E., Lunneborg, C., & Lewis, J. (1975). What does it mean to be high verbal? *Cognitive Psychology, 7,* 194–227.

Ingram, T. T. S., Mason, A. W., & Blackburn, I. (1970). A retrospective study of 82 children with reading disability. *Developmental Medicine and Child Neurology, 12,* 271–281.

Irwin, D. E., Bock, K., & Stanovich, K. E. (1982). Effects of information structure cues on visual word processing. *Journal of Verbal Learning and Verbal Behavior, 3,* 307–326.

Jackson, M. D., & McClelland, J. L. (1979). Processing determinants of reading speed. *Journal of Experimental Psychology: General, 108,* 151–181.

Jackson, M. D. (1980). Further evidence for a relationship between memory access and reading ability. *Journal of Verbal Learning and Verbal Behavior, 19,* 683–694.

Jackson, M. D., & McClelland, J. L. (1975). Sensory and cognitive determinants of reading speed. *Journal of Verbal Learning and Verbal Behavior, 19,* 565–574.

Jansky, J. J. (1979). Specificity and parameters in defining dyslexia. *Bulletin of the Orton Society, 29,* 31–38.

Johnson, D., & Myklebust, H. (1967). *Learning disabilities: Educational principles and practices.* New York: Grune & Stratton.

Joreskog, K. G., & Sorbom, D. (1978). *LISREL IV: A general computer program for estimation of linear structural equations by maximum likelihood methods.* Chicago: International Educational Services.

Just, M. A., & Carpenter, P. A. (1980). A theory of reading: From eye fixations to comprehension. *Psychological Review, 87,* 329–354.

Just, M. A., Carpenter, P. A., & Masson, M. E. J. (1982). *What eye fixations tell us about speed reading and skimming.* Carnegie-Mellon University: Eye-Lab Technical Report.

Kahneman, D. (1973). *Attention and effort.* Englewood Cliffs, NJ: Prentice-Hall.

Kail, R. V., Chi, M. T. H., Ingram, A. L., & Danner, F. W. (1977). Constructive aspects of children's reading comprehension. *Child Development, 48,* 684–688.

Kail, R. V., & Marshall, C. V. (1978). Reading skill and memory scanning. *Journal of Educational Psychology, 70,* 808–814.

Katz, L. (1977). Reading ability and single-letter orthographic redundancy. *Journal of Educational Psychology, 69,* 653–659.

Katz, L., & Feldman, L. B. (1981). Linguistic coding in word recognition: Comparisons between a deep and a shallow orthography. In A. M. Lesgold & C. A. Perfetti (Eds.), *Interactive processes in reading* (pp. 85–106). Hillsdale, NJ: Erlbaum.

Katz, R. B., Shankweiler, D., & Liberman, I. Y. (1981). Memory for item order and phonetic recoding in the beginning reader. *Journal of Experimental Child Psychology, 32*, 474–484.

Kimball, J. (1973). Seven principles of surface structure parsing in natural language. *Cognition, 2*, 15–47.

Kinsbourne, M. (1978). *Asymmetrical Function of the Brain.* Cambridge University Press.

Kintsch, W. (1974). *The representation of meaning in memory.* Hillsdale, NJ: Erlbaum.

Kintsch, W., & Keenan, J. M. (1973). Reading rate as a function of the number of propositions in the base structure of sentences. *Cognitive Psychology, 5*, 257–274.

Kintsch, W., Kozminsky, E., Streby, W. J., McKoon, G., & Keenan, J. M. (1975). Comprehension and recall of text as a function of content variables. *Journal of Verbal Learning and Verbal Behavior, 14*, 196–214.

Kintsch, W., & van Dijk, T. A. (1978). Toward a model of text comprehension and production. *Psychological Review, 85*, 363–394.

Kintsch, W., & Vipond, D. (1979). Reading comprehension and readibility in educational practice and psychological theory. In L. G. Nilsson (Ed.), *Perspectives on memory research.* Hillsdale, NJ: Erlbaum.

Kirkpatrick, J. J., & Cureton, E. E. (1949). Vocabulary item difficulty and word frequency. *Journal of Applied Psychology, 3*, 347–351.

Klare, G. R. (1974/1975). Assessing readability. *Reading Research Quarterly, 10*, 62–102.

Kleiman, G. M. (1975). Speech recoding in reading. *Journal of Verbal Learning and Verbal Behavior, 14*, 323–339.

Kliegl, R., Olson, R. K., & Davidson, B. J. (1982). Regression analyses as a tool for studying reading processes: Comments on Just and Carpenter's Eye Fixation Theory. *Memory and Cognition, 10*, 287–296.

Krashen, S. (1976). Cerebral asymmetry. In H. Avakian-Whitaker & H. Whitaker (Eds.), *Studies in neurolinguistics 2.* New York: Academic Press.

Kroll, N. E. A., Parks, T., Parkinson, S. R., Bieber, S. L., & Johnson, A. L. (1970). Short-term memory while shadowing: Recall of visually and aurally presented letters. *Journal of Experimental Psychology, 85*, 220–224.

LaBerge, D., & Samuels, S. J. (1974). Toward a theory of automatic information processing in reading. *Cognitive Psychology, 6*, 293–323.

Larkin, J. H., McDermott, J., Simon, D. P., & Simon H. A. (1980). Expert and novice performance in solving physics problems. *Science, 80*, 1335–1342.

Leahy, L. F. & Perfetti, C. A. (1984). *Reading skill differences: Short term retention and recall of phonetically confusing words.* Paper presented at the annual meeting of the American Educational Research Association, New Orleans, LA.

Leong, C. K. (1973). Reading in Chinese with reference to reading practices in Hong Kong. In J. Downing (Ed.), *Comparative Reading: Cross-national studies of behavior and processes in reading and writing.* New York: Macmillan.

Lesgold, A. M. (1983). A rationale for computer-based reading instruction. In A. C. Wilkinson (Ed.), *Classroom Computers and Cognitive Science,* 167–181. New York: Academic Press.

Lesgold, A. M., & Curtis, M. E. (1981). Learning to read words efficiently. In A. M. Lesgold & C. A. Perfetti (Eds.), *Interactive Processes in Reading.* Hillsdale, NJ: Erlbaum.

Lesgold, A. M., & Perfetti, C. A. (1978). Interactive processes in reading comprehension. *Discourse Processes, 1*, 323–336.

Lesgold, A. M., & Resnick, L. B. (1982). How reading disabilities develop: Perspectives from a longitudinal study. In J. P. Das, R. Mulcahy, & A. E. Wall (Eds.), *Theory and Research in Learning Disability.* New York: Plenum.

Lesgold, A. M., Roth, S. F., & Curtis, M. E. (1979). Foregrounding effects in discourse comprehension. *Journal of Verbal Learning and Verbal Behavior, 18,* 291–308.

Levy, B. A. (1975). Vocalization and suppression effects in sentence memory. *Journal of Verbal Learning and Verbal Behavior, 14,* 304–316.

Levy, B. A. (1978). Speech processing during reading. In A. M. Lesgold, J. W. Pellegrino, S. D. Fokkema, & R. Glaser (Eds.), *Cognitive psychology and instruction* (pp. 123–151). New York: Plenum.

Liberman, A. M., Cooper, F. S., Shankweiler, D., & Studdert-Kennedy, M. (1967). Perception of the speech code. *Psychological Review, 74,* 431–461.

Liberman, I. Y. & Shankweiler, D. (1979). Speech, the alphabet, and teaching to read. In L. B. Resnick & P. A. Weaver (Eds.). *Theory and practice of early reading Vol. 2,* Hillsdale, NJ: Erlbaum.

Liberman, I. Y., Shankweiler, D., Fischer, F. W. & Carter, B. (1974). Explicit syllable and phoneme segmentation in the young child. *Journal of Experimental Child Psychology, 18,* 201–212.

Liberman, I. Y., Shankweiler, D., Liberman, A. M., Fowler, C., & Fischer, F. W. (1977). Phonetic segmentation and recoding in the beginning reader. In A. S. Reber & D. L. Scarborough (Eds.), *Towards a psychology of reading: The proceedings of the CUNY Conference.* New York: Wiley.

Liberman, I. Y., Shankweiler, D., Orlando, C., Harris, H. S., & Berti, F. B. (1971). Letter confusion and reversals of sequence in the beginning reader: Implications for Orton's theory of developmental dyslexia. *Cortex, 7,* 127–142.

Lichtenstein, E. H., & Brewer, W. F. (1980). Memory for goal-directed events. *Cognitive Psychology, 12,* 412–445.

Lindsay, P. H., & Norman, D. A. (1977). *Human information processing: An introduction to psychology.* New York: Academic Press.

Locke, J. L., & Fehr, F. S. (1970). Subvocal rehearsal as a form of speech. *Journal of Verbal Learning and Verbal Behavior, 9,* 495–498.

Lukatela, G., Popadic, D., Ognjenovic, P., & Turvey, M. T. (1980). Lexical decision in a phonologically shallow orthography. *Memory and Cognition, 8,* 124–132.

Lundberg, I., & Torneus, M. (1978). Nonreaders' awareness of the basic relationship between spoken and written words. *Journal of Experimental Child Psychology, 25,* 404–412.

Lyon, D. R. (1977). Individual differences in immediate serial recall: A matter of mnemonics? *Cognitive Psychology, 9,* 403–411.

Mandler, J. M., & Goodman, M. (1982). On the psychological validity of story structure. *Journal of Verbal Learning and Verbal Behavior, 21,* 507–523.

Mandler, J. M., & Johnson, N. S. (1977). Remembrance of things parsed: Story structure and recall. *Cognitive Psychology, 9,* 111–151.

Mann, V. A., Liberman, I. Y., & Shankweiler, D. (1980). Children's memory for sentences and word strings in relation to reading ability. *Memory and Cognition, 8,* 329–335.

Marcel, A. (1983). Conscious and unconscious perception: Experiments on visual masking and word recognition. *Cognitive Psychology, 15,* 197–237.

Marchbanks, G., & Levin, H. (1965). Cues by which children recognize words. *Journal of Educational Psychology, 56,* 57–61.

Mark, L. S., Shankweiler, D., Liberman, I. Y., & Fowler, C. A. (1977). Phonetic recoding and reading difficulty in beginning readers. *Memory and Cognition, 5,* 623–629.

Marshall, J. C., & Newcombe, F. (1973). Patterns of paralexia: A psycholinguistic approach. *Journal of Psycholinguistic Research, 2,* 175–200.

Mason, M. (1975). Reading ability and letter search time: Effects of orthographic structures defined by single letter positional frequency. *Journal of Experimental Psychology: General, 104,* 146–166.

Massaro, D. W. (1975). *Understanding language: An information-processing analysis of speech perception, reading, and psycholinguistics.* New York: Academic Press.

Massaro, D. W. (1978). A stage model of reading and listening. *Visible Language, 12,* 3–25.

Massaro, D. W., & Taylor, G. A. (1980). Reading ability and the utilization of orthographic structure in reading. *Journal of Educational Psychology, 72,* 730–742.

Massaro, D. W., Venezky, R. L., & Taylor, G. A. (1979). Orthographic regularity, positional frequency, and visual processing of letter strings. *Journal of Experimental Psychology: General, 108,* 107–124.

Mattis, S., French, J. H., & Rapin, I. (1975). Dyslexia in children and young adults: Three independent neuropsychological syndromes. *Developmental Medicine and Child Neurology, 17,* 150–163.

McClelland, J. L. (1979). On the time relations of mental processes: An examination of systems of processes in cascade. *Psychological Review, 86,* 287–330.

McClelland, J. L., & Rumelhart, D. E. (1981). An interactive activation model of context effects in letter perception. Part 1: An account of basic findings. *Psychological Review, 88,* 357–407.

McConkie, G. W. (1982). *Eye movements and perception during reading* (Technical Report No. 229). University of Illinois at Urbana-Champaign.

McConkie, G. W., & Rayner, K. (1973). *The span of the effective stimulus during fixations in reading.* Paper presented at the annual meeting of the American Educational Research Association, New Orleans, LA.

McConkie, G. W., & Zola, D. (1981). Language constraints and the functional stimulus in reading. In A. M. Lesgold & C. A. Perfetti (Eds.), *Interactive processes in reading* (pp. 155–175). Hillsdale, NJ: Erlbaum.

McCusker, L. X., Bias, R. G., & Hillinger, M. L. (1981). Phonological recoding and reading. *Psychological Bulletin, 89,* 217–245.

McCutchen, D., & Perfetti, C. A. (1982). The visual tongue-twister effect: Phonological activation in silent reading. *Journal of Verbal Learning and Verbal Behavior, 21,* 672–687.

McGuigan, F. J. (1970). Covert oral behavior during silent performance of language tasks. *Psychological Bulletin, 74,* 309–326.

McLaughlin, G. H. (1969). Reading at "impossible" speeds. *Journal of Reading, 12,* 449–454 and 502–510.

Meichenbaum, D., & Asarnow, J. (1978). Cognitive behavioral modification and metacognitive development: Implications for the classroom. In P. Kendall & S. Hollon (Eds.), *Cognitive behavioral interventions: Theory, research, and procedures.* New York: Academic Press.

Merrill, E. C., Sperber, R. D., & McCauley, C. (1981). Differences in semantic encoding as a function of reading comprehension skill. *Memory and Cognition, 9*(6), 618–624.

Meyer, B. J. F. (1975). *The organization of prose and its effects in memory.* Amsterdam: North-Holland.

Meyer, D. E., & Ruddy, M. C. (1973, November). *Lexical memory retrieval based on graphemic and phonemic representations of printed words.* Paper presented at the meeting of the Psychonomic Society, St. Louis, MO.

Meyer, D. E., & Schvaneveldt, R. W. (1971). Facilitation in recognizing pairs of words: Evidence of a dependence between retrieval operations. *Journal of Experimental Psychology, 90,* 227–234.

Meyer, D. E., Schvaneveldt, R., & Ruddy, M. (1974). Functions of graphemic and phonemic codes in visual word recognition. *Memory and Cognition, 2,* 309–321.

Mitchell, D. C., & Green, D. W. (1978). The effects of context and content on immediate processing in reading. *Quarterly Journal of Experimental Psychology, 30,* 609–636.

Moore, M. J., Kagan, J., Sahl, M., & Grant, S. (1982). Cognitive profiles in reading disability. *Genetic Psychology Monographs, 105,* 41–93.

Morais, J., Cary, L., Alegria, J., & Bertelson, P. (1979). Does awareness of speech as a sequence of phones arise spontaneously? *Cognition, 7,* 323–331.

Morrison, F. J. (1980, November). *Reading disability: Toward a reconceptualization.* Paper presented at the annual meeting of the Psychonomic Society, St. Louis, MO.

Morton, J. (1964). The effects of context on the visual duration threshold for words. *British Journal of Psychology, 85,* 165–180.

Morton, J. (1969). Interaction of information in word recognition. *Psychological Review, 76,* 165–178.

Naidoo, S. (1972). *Specific dyslexia.* New York: Wiley, Halstead Press.

Naish, P. (1980). The effects of graphemic and phonemic similarity between targets and masks in a backward visual masking paradigm. *Quarterly Journal of Experimental Psychology, 32,* 57–68.

Navon, D., & Shimron, J. (1981). Does word naming involve grapheme-to-phoneme translation? Evidence from Hebriew. *Journal of Verbal Learning and Verbal Behavior, 20,* 97–109.

Neely, J. H. (1977). Semantic priming and retrieval from lexical memory: The roles of inhibitionless spreading activation and limited-capacity attention. *Journal of Experimental Psychology: General, 106,* 1–66.

Neisser, U. (1976). General, academic, and artificial intelligence. In L. B. Resnick (Ed.), *The nature of intelligence.* Hillsdale, NJ: Erlbaum.

Newell, A., & Simon, H. A. (1972). *Human problem solving.* Englewood Cliffs, NJ: Prentice-Hall.

Norman, D. A. (1976). *Memory and attention.* New York: Wiley.

Norman, D. A., & Bobrow, D. (1979). Descriptions: An intermediate stage in memory retrieval. *Cognitive Psychology, 11,* 107–123.

O'Donnell, R. C. (1974). Syntactic differences between speech and writing. *American Speech, 49,* 102–110.

Olson, D. R. (1977). From utterance to text: The bias of language in speech and writing. *Harvard Educational Review, 47,* 257–281.

Olson, R. K., Davidson, B. J., Kliegl, R., & Davies, S. (1984). Development of phonological memory in disabled and normal readers. *Journal of Experimental Child Psychology, 37,* 187–206.

Olson, R. K., Kliegl, R., & Davidson, B. J. (1983a). Eye movements in reading disability. In K. Rayner (Ed.), *Eye movements in reading: Perceptual and language processes.* New York: Academic Press.

Olson, R. K., Kliegl, R., & Davidson, B. J. (1983b). Dyslexic and normal children's tracking eye movements. *Journal of Experimental Psychology: Human Perception and Performance, 9,* 816–825.

Olson, R. K., Kliegl, R., Davidson, B. J., & Foltz, G. (1984). Individual and developmental differences in reading disability. In T. G. Waller (Ed.), *Reading research: Advances in theory and practice.* New York: Academic Press.

Omanson, R. C. (1982a). An analysis of narratives: Identifying central, supportive, and distracting content. *Discourse Processes, 5,* 195–224.

Omanson, R. C. (1982b). The relation between centrality and story category variation. *Journal of Verbal Learning and Verbal Behavior, 21,* 326–337.

Omanson, R. C., Beck, I. L., McKeown, M. G., & Perfetti, C. A. (In press). Comprehension of texts with unfamiliar versus unfamiliar taught words: An assessment of models. *Journal of Educational Psychology.*

Omanson, R. C., & Malamut, S. R. (1980, November). *The effects of supportive and distracting content on the recall of central content.* Paper presented at the annual meeting of the Psychonomic Society, St. Louis, MO.

Orton, S. T. (1925). Word-blindness in school children. *Archives of Neurology and Psychiatry, 14,* 581–615.

Orton, S. T. (1937). *Reading, writing, and speech problems in children.* New York: Norton.

Palinscar, A. S., & Brown, A. L. (1984). Reciprocal teaching of comprehension-fostering and comprehension-monitoring activities. *Cognition and Instruction, 1*(2), 117–175.

Paris, S. G., & Lindaur, B. K. (1976). The role of inference in children's comprehension & memory for sentences. *Cognitive Psychology, 8,* 217–227.

Patterson, K., & Marcel, A. (1977). Aphasia, dyslexia, and the phonological code of written words. *Quarterly Journal of Experimental Psychology, 29,* 307–318.

Pavlidis, G. T. (1981). Do eye movements hold the key to dyslexia? *Neuropsychologia, 19,* 57–64.

Perfetti, C. A. (1969). Lexical density and phrase structure depth as variables in sentence retention. *Journal of Verbal Learning and Verbal Behavior, 8,* 719–724.

Perfetti, C. A. (1977). Language comprehension and fast decoding: Some psycholinguistic prerequisites for skilled reading comprehension. In J. T. Guthrie (Ed.), *Cognition, Curriculum, and Comprehension,* 20–41.

Perfetti, C. A. (1983). Individual differences in verbal processes. In R. Dillon & R. R. Schmeck (Eds.), *Individual differences in cognition.* New York: Academic Press.

Perfetti, C. A. Language, speech, and print: Some asymmetries in the acquisition of literacy. In R. Horowitz & S. J. Samuels (Eds.), *Comprehending Oral and Written Language.* New York: Academic Press. In press.

Perfetti, C. A., Beck, I. L., & Hughes, C. (1981, April). *Phonemic knowledge and learning to read.* Paper presented at the Society for Research in Child Development. Boston, MA.

Perfetti, C. A., & Bell, L. C. (1983). *Reading skill and the use of structure in letter search.* Unpublished manuscript, University of Pittsburgh.

Perfetti, C. A., Finger, E., & Hogaboam, T. W. (1978). Sources of vocalization latency differences between skilled and less skilled young readers. *Journal of Educational Psychology, 70*(5), 730–739.

Perfetti, C. A., & Goldman, S. R. (1976). Discourse memory and reading comprehension skill. *Journal of Verbal Learning and Verbal Behavior, 14,* 33–42.

Perfetti, C. A., Goldman, S. R., & Hogaboam, T. W. (1979). Reading skill and the identification of words in discourse context. *Memory and Cognition, 7,* 273–282.

Perfetti, C. A., & Hogaboam, T. W. (1975). The relationship between single word decoding and reading comprehension skill. *Journal of Educational Psychology, 67,* 461–469.

Perfetti, C. A., & Lesgold, A. M. (1977). Discourse comprehension and sources of individual differences. In M. A. Just & P. A. Carpenter (Eds.), *Cognitive processes in comprehension* (pp. 141–183). Hillsdale, NJ: Erlbaum.

Perfetti, C. A., & Lesgold, A. M. (1979). Coding and comprehension in skilled reading and implications for reading instruction. In L. B. Resnick & P. A. Weaver (Eds.), *Theory and practice of early reading, Vol. 1.* Hillsdale, NJ: Erlbaum.

Perfetti, C. A., & McCutchen, D. (1982). Speech processes in reading. In N. Lass (Ed.), *Speech and language: Advances in baisc research and practice* (Vol. 7) (pp. 237–269). New York: Academic Press.

Perfetti, C. A., Riley, M., & Greeno, J. G. (1978, November). *Comprehension and computation: The role of sentence encoding in verbal arithmetic.* Paper presented at the annual meeting of the Psychonomic Society, San Antonio, TX.

Perfetti, C. A., & Roth, S. F. (1981). Some of the interactive processes in reading and their role in reading skill. In A. M. Lesgold & C. A. Perfetti (Eds.), *Interactive processes in reading* (pp. 269–297). Hillsdale, NJ: Erlbaum.

Petrick, S., & Potter, M. C. (1979, November). *RSVP sentences and word lists: Representation of meaning and sound.* Paper presented at the annual meeting of the Psychonomic Society, Phoenix, AZ.

Piggins, W. R., & Barron, R. W. (1982). *Why learning to read aloud more rapidly does not improve comprehension: Testing the decoding sufficiency hypothesis.* Paper presented at the annual meeting of the American Educational Research Association, New York, NY.

Pirozzolo, F. J., & Rayner, K. (1978). The normal control of eye movements in acquired and developmental reading disorders. In H. Avakian-Whitaker & H. A. Whitaker (Eds.), *Advances in neurolinguistics and psycholinguistics.* New York: Academic Press.

Posner, M. I. (1969). Abstraction and the process of recognition. In G. H. Bower & J. T. Spence (Eds.), *Psychology of learning and motivation* (Vol. 3). New York: Academic Press.

Posner, M. I., & Mitchell, R. (1967). Chronometric analysis of classification. *Psychological Review, 74,* 392–409.

Posner, M. I., & Snyder, C. R. R. (1975). Attention and cognitive control. In R. Solso (Ed.), *Information processing and cognition: The Loyola symposium.* Hillsdale, NJ: Erlbaum.

Rayner, K. (1975). The perceptual span and peripheral cues in reading. *Cognitive Psychology, 7,* 65–81.

Rayner, K. (1976). Developmental changes in word recognition strategies. *Journal of Educational Psychology, 68,* 323–334.

Rayner, K. (1977). Visual attention in reading: Eye movements reflect cognitive processes. *Memory and Cognition, 5(4),* 443–448.

Rayner, K. (1978). Eye movements in reading and information processing. *Psychological Bulletin, 85,* 618–660.

Rayner, K., & Hagelberg, E. M. (1975). Word recognition cues for beginning and skilled readers. *Journal of Experimental Child Psychology, 20,* 444–455.

Rayner, K., & McConkie, G. W. (1976). What guides a reader's eye movements? *Vision Research, 16,* 829–837.

Rayner, K, McConkie, G. W., & Zola, D. (1980). Integrating information across eye movements. *Cognitive Psychology, 12,* 206–226.

Rayner, K., Well, A. D., Pollatsek, A., & Bertera, J. H. (1982). The availability of useful information to the right of fixation in reading. *Perception and Psychophysics, 31,* 537–550.

Read, C. (1971). Pre-school children's knowledge of English phonology. *Harvard Educational Review, 41,* 1–34.

Richardson, G. (1982). *The perception of orientation and form recognition under spatial transformation in retarded and normal readers.* Unpublished manuscript, Cambridge University.

Rosner, J. (1973). Language arts and arithmetic achievement and specifically related perceptual skills. *American Educational Research Journal, 10,* 59–68.

Rozin, P., Bressman, B., & Taft, M. (1974). Do children understand the basic relationship between speech and writing? The Mow-Motorcycle test. *Journal of Reading Behavior, 6,* 327–334.

Rozin, P., & Gleitman, L. R. (1977). The structure and acquisition of reading II: The reading process and the acquisition of the alphabetic principle. In A. S. Reber & D. L. Scarborough (Eds.), *Toward a psychology of reading: Proceedings of the CUNY conference* (pp. 55–141). Hillsdale, NJ: Erlbaum.

Rubenstein, H., Lewis, S. S., & Rubenstein, M. A. (1971). Evidence for phonemic reading in visual word recognition. *Journal of Verbal Learning and Verbal Behavior, 10*, 645–657.

Rubin, A. (1980). A theoretical taxonomy of the differences between oral and written language. In B. J. Spiro, B. C. Bruce, & W. F. Brewer (Eds.), *Theoretical issues in reading comprehension*. Hillsdale, NJ: Erlbaum.

Rumelhart, D. E. (1975). Notes on a schema for stories. In D. Bobrow & A. Collins (Eds.), *Representation and understanding: Studies in cognitive science*. New York: Academic Press.

Rumelhart, D. E. (1977). Toward an interactive model of reading. In S. Dornic & P. M. A. Rabbitt (Eds.), *Attention and performance VI*. Hillsdale, NJ: Erlbaum.

Rumelhart, D. E., & McClelland, J. L. (1981). Interactive processing through spreading activation. In A. M. Lesgold & C. A. Perfetti (Eds.), *Interactive processes in reading* (pp. 37–60). Hillsdale, NJ: Erlbaum.

Rumelhart, D. E., & McClelland, J. L. (1982). An interactive activation model of context effects in letter perception. Part 2: The contextual enhancement effect and some tests and extensions of the model. *Psychological Review, 89*, 60–94.

Rumelhart, D. E., & Ortony, A. (1977). The representation of knowledge in memory. In R. C. Anderson, R. J. Spiro, & W. E. Montague (Eds.), *Schooling and the acquisition of knowledge*. Hillsdale, NJ: Erlbaum.

Ryan, E. B. (1978). Metalinguistic development and reading. In F. B. Murray (Ed.), *The development of the reading process*. Newark, DE: International Reading Association.

Ryan, E. B. (1982). Identifying and remediating failures in reading comprehension: Toward an instructional approach for poor comprehenders. In G. E. MacKinnon & T. G. Waller (Eds.), *Advances in Reading Research* (Vol. 3). New York: Academic Press.

Samuels, S. J. (1979). The method of repeated readings. *The Reading Teacher, 32*(4) 403–408.

Satz, P. (1976). Cerebral dominance and reading disability: An old problem revisited. In R. M. Knights & D. J. Bakker (Eds.), *The neuropsychology of learning disorders: Theoretical approaches*. Baltimore: University Park Press.

Satz, P., & Morris, R. (1981). Learning disability subtypes: A review. In F. J. Pirozzolo & M. C. Wittrock (Eds.), *Neuropsychological and cognitive processes in reading*. New York: Academic Press.

Schank, R. C. (1982). *Reading and understanding: Teaching from the perspective of artificial intelligence*. Hillsdale, NJ: Erlbaum.

Schank, R. C., & Abelson, R. P. (1977). *Scripts, plans, goals, and understanding: An inquiry into human knowledge structures*. Hillsdale, NJ: Erlbaum.

Schneider, W., & Shiffrin, R. M. (1977). Controlled and automatic human information processing. I: Detection, search, and attention. *Psychological Review, 84*, 1–66.

Schuberth, R. E., & Eimas, P. D. (1977). Effects of context on the classification of words and nonwords. *Journal of experimental psychology: Human perception and performance, 3*, 27–36.

Seegers, G., & Feenstra, J. (1982). *Een onderzoek naar leesvaardigheidsverschillen tussen leerlingen uit klas 3 en klas 5 van het gewoon lager onderwijs* (Report No. 2). Interdisciplinaire Studierichting Onderwijskunde Katholieke Universiteit Nijmegen (Netherlands).

Shallice, R., & Warrington, E. (1975). Word recognition in a phonemic dyslexic patient. *Quarterly Journal of Experimental Psychology, 27*, 148–160.

Shankweiler, D., & Liberman, I. Y. (1972). Misreading: A search for causes. In J. F. Kavanagh & I. G. Mattingly (Eds.), *Language by ear and by eye* (pp. 293–317). Cambridge, MA: MIT Press.

Shankweiler, D., Liberman, I. Y., Mark, L. S., Fowler, C. A., & Fischer, F. W. (1979). The speech code and learning to read. *Journal of Experimental Psychology: Human Learning and Memory, 5,* 531–545.

Shiffrin, R. M., & Schneider, W. (1977). Controlled and automatic human information processing. II: Perceptual learning, automatic attending, and a general theory. *Psychological Review, 84,* 127–199.

Short, E. J., & Ryan, E. B. (1982). *Remediating poor readers' comprehension failures with a story grammar strategy.* Paper presented at the annual meeting of the American Educational Research Association, New York, NY.

Simpson, G. B. (1981). Meaning dominance and semantic context in the processing of lexical ambiguity. *Journal of Verbal Learning and Verbal Behavior, 20,* 120–136.

Slowiaczek, M. L., & Clifton, C. (1980). Subvocalization and reading for meaning. *Journal of Verbal Learning and Verbal Behavior, 19,* 573–582.

Smiley, S. S., Oakley, D. D., Worthen, D., Campione, J. C., & Brown, A. L. (1977). Recall of thematically relevant material by adolescent good and poor readers as a function of written and oral presentation. *Journal of Educational Psychology, 69,* 881–887.

Smith, E. E., Shoben, E. J., & Rips, L. J. (1974). Structure and process in semantic memory: A featural model for semantic decisions. *Psychological Review, 81,* 214–241.

Smith, F. (1971). *Understanding reading.* New York: Holt, Rinehart and Winston.

Smith, F. (1973). *Psycholinguistics and reading.* New York: Holt, Rinehart and Winston.

Sokolov, A. N. (1972). *Inner speech and thought* (D. Lindsley, trans.). New York: Plenum.

Sperling, G. (1960). The information available in brief visual presentations. *Psychological Monographs, 74,* 20 (Whole No. 498).

Sperry, R. W. (1964). The great cerebral commissure. *Scientific American, 210,* 42–52.

Spilich, G. J., Vesonder, G. T., Chiesi, H. L., & Voss, J. F. (1979) Text-processing of domain-related information for individuals with high and low domain knowledge. *Journal of Verbal Learning and Verbal Behavior, 18,* 275–290.

Spiro, R. J. (1980). Constructive processes in prose comprehension and recall. In R. J. Spiro, B. C. Bruce, & W. F. Brewer (Eds.), *Theoretical issues in reading comprehension,* (pp. 245–278). Hillsdale, NJ: Erlbaum.

Spring, C., & Capps, C. (1974). Encoding speed, rehearsal, and probed recall of dyslexic boys. *Journal of Educational Psychology, 66,* 780–786.

Stanley, G., Smith, G. A., & Howell, E. A. (1983). Eye movements and sequential tracking in dyslexic and control children. *British Journal of Psychology, 74,* 181–187.

Stanovich, K. E. (1980). Toward an interactive-compensatory model of individual differences in the development of reading fluency. *Reading Research Quarterly, 16,* 32–71.

Stanovich, K. E. (1981). Attentional and automatic context effects in reading. In A. M. Lesgold & C. A. Perfetti (Eds.), *Interactive processes in reading* (pp. 241–267). Hillsdale, NJ: Erlbaum.

Stanovich, K. E., & Bauer, D. (1978). Experiments on the spelling-to-sound regularity effect in word recognition. *Memory and Cognition, 6,* 410–415.

Stanovich, K. E., & West, R. F. (1981). The effect of sentence context on on-going word recognition: Tests of a two-process theory. *Journal of Experimental Psychology: Human Perception and Performance, 7,* 658–672.

Stanovich, K. E., West, R. F., & Feeman, D. J. (1981). A longitudinal study of sentence context effects in second-grade children: Tests of an interactive-compensatory model. *Journal of Experimental Child Psychology, 32,* 185–199.

Stein, N. L., & Glenn, C. G. (1979). An analysis of story comprehension in elementary school children. In R. Freedle (Ed.), *Advances in discourse process-ing 2: New Directions in discourse processing.* Norwood, NJ: Ablex.

Steiner, R., Wiener, M., & Cromer, W. (1971). Comprehension training and iden-tification for poor and good readers. *Journal of Educational Psychology, 62*(6), 506–513.

Sternberg, R. J. (1981). Toward a unified componential theory of intelligence. I: Fluid ability. In M. Friedman, J. Das, and N. O'Connor (Eds.), *Intelligence and Learning.* New York: Plenum.

Sternberg, R. J., Powell, J. S., & Kaye, D. B. (1983). Teaching vocabulary-building skills: A contextual approach. In A. C. Wilkinson (Ed.), *Classroom comput-ers and cognitive science.* New York: Academic Press.

Sticht, T. G. (1977). Comprehending reading at work. In M. A. Just & P. A. Car-penter (Eds.), *Cognitive processes in comprehension.* Hillsdale, NJ: Erlbaum.

Sticht, T. G. (1979). Applications of the Audread model to reading evaluation and instruction. In L. B. Resnick & P. A. Weaver (Eds.), *Theory and practice of early reading, Vol. 1.* Hillsdale, NJ: Erlbaum.

Swinney, D. A. (1979). Lexical access during sentence comprehension: Recon-sideration of context effects. *Journal of Verbal Learning and Verbal Behav-ior, 18,* 645–659.

Tanenhaus, M. K., Lieman, J. M., & Seidenberg, M. T. (1979). Evidence for mul-tiple stages in the processing of ambiguous words in syntactic contexts. *Jour-nal of Verbal Learning and Verbal Behavior, 18,* 427–440.

Tannen, D. (1982). Oral and literate strategies in spoken and written narratives. *Language, 58,* 1–21.

Taylor, S. E. (1962). An evaluation of forty-one trainees who had recently com-pleted the "Reading Dynamics" program. *Eleventh yearbook of the national reading conference,* 41–55.

Thomas, E. L. (1962). Eye movements in speed reading. In R. G. Stauffer (Ed.), *Speed reading: Practices and procedures* (Vol. 10) (pp. 104–114). Newark, DE: University of Delaware, Reading Study Center.

Thorndyke, P. W. (1977). Cognitive structures in comprehension and memory of narrative discourse. *Cognitive Psychology, 9,* 77–110.

Thorndyke, P. W., & Hayes-Roth, B. (1979). The use of schemata in the acquisi-tion and transfer of knowledge. *Cognitive Psychology, 11,* 82–106.

Tinker, M. A. (1958). Recent studies of eye movements in reading. *Psychological Bulletin, 58,* 215–231.

Torgesen, J. K. (1978). Performance of reading disabled children on serial mem-ory tasks: A selective review of recent research. *Reading Research Quarterly, 14,* 57–87.

Torgesen, J. K. (1982). The use of rationally defined subgroups in research on learning disabilities. In J. P. Das, R. F. Mulcahy, & A. E. Wall (Eds.), *Theory and research in learning disabilities.* New York: Plenum.

Torgesen, J. K., & Goldman, T. (1977). Verbal rehearsal and short term memory in reading-disabled children. *Child Development, 48,* 56–60.

Torrey, J. W. (1979). Reading that comes naturally: The early reader. In T. G. Waller & G. E. MacKinnon (Eds.), *Reading research: Advances in theory and prac-tice* (Vol. 1). New York: Academic Press.

Townsend, M. A. R. (1982). Flexibility of schema shifting in good and poor read-ers. *Journal of Reading Behavior, 14,* 169–178.

Trabasso, T., Rollins, H., & Shaughnessy, E. (1971). Storage and verification stages in processing concepts. *Cognitive Psychology, 2,* 239–289.

Trieman, R. A. (1976). *Children's ability to segment speech into syllables and phonemes as related to their reading ability.* Unpublished manuscript, Yale University.

Trieman, R. A., & Baron, J. (1982). Segmental analysis ability: Development and relation to reading ability. In T. G. Waller & G. E. MacKinnon (Eds.), *Reading research: Advances in theory and practice* (Vol. 3). New York: Academic Press.

Trieman, R. A., & Breaux, A. M. (1982). Common phoneme and overall similarity relations among spoken syllables. *Journal of Psycholinguistic Research, 11,* 569–597.

Tulving, E., & Gold, C. (1963). Stimulus information and contextual information as determinants of tachistoscopic recognition of words. *Journal of Experimental Psychology, 66,* 319–327.

Turvey, M. T. (1973). On peripheral and central processes in vision: Inferences from an information-processing analysis of masking with patterned stimuli. *Psychological Review, 80,* 1–52.

Valtin, R. (1973). *Report of research on dyslexia in children.* Paper presented at the International Reading Association, Denver, CO. (ERIC document ED 079 713).

van den Bos, K. P. (1980). Cognitive abilities and learning disabilities. *Bulletin of the Orton Society, 30,* 94–111.

van den Bos, K. P. (1982). *Letter span, scanning, and code matching in dyslexic subgroups.* Paper presented at the thirty-third annual conference of the Orton Society, Baltimore, MD.

Vellutino, F. R. (1979). *Dyslexia: Theory and research.* Cambridge, MA: MIT Press.

Vellutino, F. R. (1980). Dyslexia—perceptual deficiency or perceptual inefficiency? In J. F. Kavanagh & R. L. Venezky (Eds.), *Orthography, reading, and dyslexia.* Baltimore, MD: University Park Press.

Vellutino, F. R., Smith, H., Steger, J. A., & Kaman, M. (1975). Reading disability: Age differences and the perceptual deficit hypothesis. *Child Development, 46,* 487–493.

Vellutino, F. R., Steger, J. A., & Kandel, G. (1972). Reading disability: An investigation of the perceptual deficit hypothesis. *Cortex, 8,* 106–118.

Venezky, R. L. (1967). English orthography: Its graphical structure and its relation to sound. *Reading Research Quarterly, 2,* 75–106.

Venezky, R. L. (1970). *The structure of English orthography.* The Hague: Mouton.

Venezky, R. L. (1978). Reading acquisition: The occult and the obscure. In F. B. Murray & J. J. Pikulski (Eds.), *The acquisition of reading.* Baltimore, MD: University Park Press.

Vesonder, G. T. (1979). *The role of knowledge in the processing of experimental reports.* Unpublished doctoral dissertation, University of Pittsburgh.

Vogel, S. A. (1975). *Syntactic abilities in normal and dyslexic children.* Baltimore, MD: University Park Press.

Vonk, W., & Noordman, L. G. M. (1982). *Making inference during text understanding.* Paper presented at the Conference on Language, Reasoning, and Inference, Edinburgh.

Warren, W. H., Nicholas, D. W., & Trabasso, T. (1979). Event chains and inferences in understanding narratives. In R. Freedle (Ed.), *Advances in discourse processes* (Vol. 2). Norwood, NJ: Ablex.

Waters, G. S. (1981). *Interference effects on reading: Implications for phonological recoding.* Unpublished doctoral dissertation, Concordia University.

Waugh, N. C., & Norman, D. A. (1965). Primary memory. *Psychological Review, 72,* 89–104.

Weaver, P. A., & Dickinson, D. K. (1982). Scratching below the surface structure: Exploring the usefulness of story grammars. *Discourse Processes, 5,* 225–243.

Weber, R. (1970). A linguistic analysis of first grade reading errors. *Reading Research Quarterly, 5,* 427–451.

West, R. F., & Stanovich, K. E. (1978). Automatic contextual facilitation in readers of three ages. *Child Development, 49,* 717–727.

Wilkinson, A. C. (Ed.) (1983). *Classroom computers and cognitive science.* New York: Academic Press.

Wilkinson, A. C., Epstein, W., Glenberg, A. M., & Morse, E. (1980). *The illusion of knowing in studying texts.* Paper presented at the annual meeting of the Psychonomic Society, St. Louis, MO.

Wilkinson, A. C., & Patterson, J. (1983). Issues at the interface of theory and practice. In A. C. Wilkinson (Ed.), *Classroom computers and cognitive science*
(pp. 3–13). New York: Academic Press.

Williams, J. P. (1979). Reading instruction today. *American Psychologist, 34,* 917–
922.

Williams, J. P., Blumberg, E. L., & Williams, D. V. (1970). Cues used in visual
word recognition. *Journal of Educational Psychology, 61,* 310–315.

Willows, D. M., Borwick, D., & Hayvren, M. (1981). The content of school readers. In G. E. MacKinnon & T. G. Waller (Eds.), *Reading research: Advances
in theory and practice* (Vol. 2). New York: Academic Press.

Wood, C. C., Goff, W. R., & Day, R. S. (1971). Auditory evoked potentials during
speech perception. *Science, 173,* 1248–1251.

Zola, D. (1979). *The perception of words in reading.* Paper presented at the annual meeting of the Psychonomic Society, Phoenix, AZ.

Zurif, E., & Carson, G. (1970). Dyslexia in relation to cerebral dominance and
temporal analysis. *Neuropsychologia, 8,* 351–362.

AUTHOR INDEX

SUBJECT INDEX